CONTROL AND GAME THEORETIC METHODS FOR CYBER-PHYSICAL SECURITY

Emerging Methodologies and Applications in Modelling, Identification and Control

CONTROL AND GAME THEORETIC METHODS FOR CYBER-PHYSICAL SECURITY

ARIS KANELLOPOULOS
Division of Information Science and Engineering
KTH Royal Institute of Technology
Stockholm, Sweden

LIJING ZHAI
Daniel Guggenheim School of Aerospace Engineering
Georgia Institute of Technology
Atlanta, GA, United States

FILIPPOS FOTIADIS
Daniel Guggenheim School of Aerospace Engineering
Georgia Institute of Technology
Atlanta, GA, United States

KYRIAKOS G. VAMVOUDAKIS
Daniel Guggenheim School of Aerospace Engineering
Georgia Institute of Technology
Atlanta, GA, United States

Series Editor
QUAN MIN ZHU

ELSEVIER

ACADEMIC PRESS
An imprint of Elsevier

Academic Press is an imprint of Elsevier
125 London Wall, London EC2Y 5AS, United Kingdom
525 B Street, Suite 1650, San Diego, CA 92101, United States
50 Hampshire Street, 5th Floor, Cambridge, MA 02139, United States

ISBN: 978-0-443-15408-9

For information on all Academic Press publications
visit our website at https://www.elsevier.com/books-and-journals

Publisher: Mara Conner
Acquisitions Editor: Sophie Harrison
Editorial Project Manager: Namrata Lama
Production Project Manager: Fizza Fathima
Cover Designer: Mark Rogers

Typeset by VTeX

This book is dedicated to my parents, Sofia and Antonis.

Aris Kanellopoulos

This book is dedicated to my parents Hongzhi and Haichao, and my cutie pie Sunny.

Lijing Zhai

This book is dedicated to my parents, Despoina and Kosmas.

Filippos Fotiadis

This book is dedicated to my parents Evgenia and George for always loving and supporting me. It is also dedicated to my mentors through the years, Frank L. Lewis, who gave me the initial passion for the field and taught me how to come up with new ideas, and João P. Hespanha, who taught me how to think outside the box while also being a perfectionist.

Kyriakos G. Vamvoudakis

Contents

Biographies

Aris Kanellopoulos

Aris Kanellopoulos was born in Athens, Greece. He received his diploma equivalent to a Master of Science in Mechanical Engineering from the National Technical University of Athens, Greece, in 2017. He studied at the Kevin T. Crofton Department of Aerospace and Ocean Engineering at Virginia Tech and was awarded a Ph.D. in Aerospace Engineering at the Georgia Institute of Technology. In 2022, he was a Research Engineer within Professor Kyriakos G. Vamvoudakis' group. He is currently a Postdoctoral Researcher with the Division of Information Science and Engineering at the Royal Institute of Technology, Sweden.

Lijing Zhai

Lijing Zhai was born in Luoyang, China. She received a Master of Science in Power Engineering and Engineering Thermophysics from Beihang University, Beijing, China, in 2018. She was awarded the Ph.D. degree in Aerospace Engineering at the Georgia Institute of Technology in 2023. Her research interests include cyber-physical security, reinforcement learning, and data-driven control.

Filippos Fotiadis

Filippos Fotiadis was born in Thessaloniki, Greece. He received the Diploma (joint B.Sc. and M.Sc. degree) in Electrical and Computer Engineering from the Aristotle University of Thessaloniki, Greece, in 2018. He is currently pursuing the Ph.D. degree in Aerospace Engineering at the Georgia Institute of Technology, where he also obtained the M.Sc. degree in Aerospace Engineering in 2022 and the M.Sc. degree in Mathematics in 2023. His research interests include optimal and learning-based control, game theory, and their applications to cyber-physical security, as well as prescribed performance control.

Kyriakos G. Vamvoudakis

Kyriakos G. Vamvoudakis was born in Athens, Greece. He received the Diploma (a 5-year degree, equivalent to the M.Sc. degree) in Electronic and Computer Engineering from the Technical University of Crete, Greece, in 2006 with highest honors. After moving to the United States of America, he studied at The University of Texas at Arlington with Frank L. Lewis as his advisor, and he received his M.S. and Ph.D. in Electrical Engineering in 2008 and 2011, respectively. During the period from 2012 to 2016, he was project research scientist at the Center for Control, Dynamical Systems and Computation at the University of California, Santa Barbara. He was an Assistant Professor at the Kevin T. Crofton Department of Aerospace and Ocean Engineering at Virginia Tech until 2018. He currently serves as the Dutton-Ducoffe Endowed Professor at The Daniel Guggenheim School of Aerospace Engineering at Georgia Tech. He holds a secondary appointment in the School of Electrical and Computer Engineering. His expertise is in reinforcement learning, control theory, game theory, cyber-physical security, bounded rationality, and safe/assured autonomy. Dr. Vamvoudakis is the recipient of a 2019 ARO YIP award, a 2018 NSF CAREER award, a 2018 DoD Minerva Research Initiative Award, a 2021 GT Chapter Sigma Xi Young Faculty Award, and his work has been recognized with best paper nominations and several international awards including the 2016 International Neural Network Society Young Investigator (INNS) Award, the Best Paper Award for Autonomous/Unmanned Vehicles at the 27th Army Science Conference in 2010, and the Best Researcher Award from the Automation and Robotics Research Institute in 2011. He currently is an Associate Editor of many journals. He is also a registered Electrical/Computer engineer (PE), a member of the Technical Chamber of Greece, an Associate Fellow of AIAA, and a Senior Member of IEEE.

CHAPTER ONE

Introduction

1.1. Cyber-physical systems and security

Cyber-physical systems (CPS) are systems of high heterogeneity and complexity, comprising multiple digital and physical components that interact with one another through a variety of communication channels and computational elements (Rajkumar et al., 2010). CPS have been at the forefront of research endeavors by a multitude of research communities with vastly different viewpoints. Computer scientists approach the design of CPS with great focus given to the interface between the computational elements of such systems, to the embedded controllers that are connected through the internet (Wolf, 2009), and to the communication protocols that dictate the operation of the digital-to-physical interface (Huh et al., 2017). Simultaneously, software engineers have been keen on analyzing how the decentralized and complex nature of CPS can be tackled to guarantee smooth coordination of all the subsystems (Lee, 2008). Following a more mathematically focused direction, control theorists have been exploring issues of CPS in the context of decentralized control (Pasqualetti et al., 2013), robust methods (Jin et al., 2017; Sun et al., 2019), and optimization techniques (Cao et al., 2012).

The extensive attention that CPS have been receiving from researchers can be attributed to their ubiquity in modern society. Such systems can be found in several areas ranging from military to civilian applications. In healthcare and medicine (Lee and Sokolsky, 2010), medical CPS are employed to augment medical devices with networking capabilities that enable real-time monitoring of the patient and allows autonomously handling routine situations. Smart grids (Liu et al., 2011; Mo et al., 2012) will act as the next-generation providers of electricity in networks that are self-monitoring and self-configurable. Analyzing those systems as CPS offers a unique understanding of their operation and supports the development of novel algorithms to optimize their operation. One of the major fields where CPS are found is the field of transportation (Kim et al., 2013). From unmanned vehicles (Jin et al., 2019) to autonomous delivery systems and supply chains (Flämig, 2016), CPS will need to be integrated into the extremely human-centric environment of the traffic network.

Control and Game Theoretic Methods for Cyber-Physical Security
https://doi.org/10.1016/B978-0-44-315408-9.00007-5

Due to the complex and often large-scale nature of CPS, they are an enticing target for adversaries that may want to create confusion, disruption, and performance deterioration through a cyber-physical attack. Numerous attacks on CPS have been reported with a few relevant examples including the German steel mill attack (Lee et al., 2014), the 2015 Ukrainian blackout (Liang et al., 2016), and the Saudi Aramco incident (Bronk and Tikk-Ringas, 2013), going as far back as the Stuxnet virus, a malicious computer worm targeting programmable logic controllers (Farwell and Rohozinski, 2011) or the attack on the Maroochy water services in Australia (Slay and Miller, 2007). Thus it becomes apparent that protecting CPS in both physical and digital domains should not be a long-term goal against potential, futuristic attacks, but a pressing issue that must be confronted immediately. This, in fact, generated an increasing demand for secure methods that can guarantee the integrity and normal operation of CPS under adversarial attacks (Cardenas et al., 2008). A fault taxonomy of attacks, faults, and detection mechanisms is provided in Zhai et al. (2022), which assists a CPS system designer or operator in selecting a detection mechanism for identified attack or fault scenarios.

The issue of securing systems has been of the utmost importance in nearly every facet of human activity. Recently, the area of cyber-security has been pushed forward via the development of a variety of mechanisms, including cryptographic techniques, network resilience methods, and ad hoc solutions to various attack classes recognized by the computer science community. However, the advent of more complex dynamical and adaptive systems has led to the need for more complete, mathematically supported and system-oriented approaches. This is further exacerbated by the increasing penetration of autonomous learning-based components in both military and civilian applications. Adding to this complexity, autonomous agents will be required to operate in closer proximity to humans, being vulnerable to the full unpredictability of the dynamical nature of their behavior. This is especially true when security issues are considered, since the demonstrated imagination and creativity of hackers is a clear indicator that naive, static, equilibrium-oriented models of attack behavior — ones that might consider worst-case scenarios, or specific, constant adversarial choices — are bound to be insufficient once complex systems are further used in everyday life.

In their majority, security mechanisms, both for computer systems and CPS, are reactive in nature. Various fields of research have thrived in their pursuit of approaches to shielding systems against attackers or in the development of detection mechanisms. These approaches, however, fail to consider zero-day attacks to systems and disregard the dynamical nature of the abilities of cyber-attackers. This creates an asymmetric situation, in which the defenders employ static tools that are rigid in their adaptability, whereas the attackers can plan and perform extensive reconnaissance over lengthy periods of time while launching the actual attack in short bursts.

1.2. The viewpoint of control theorists

The work of Cardenas et al. (2008) was one of the first published manuscripts that questioned the adequacy of the security approaches operating only in the computational layer, such as encryption algorithms. Therefore extensive research has been conducted on the behavior and security of complex CPS from a control-theoretic standpoint (Pajic et al., 2017; Pasqualetti et al., 2015; Satchidanandan and Kumar, 2017). Furthermore, by leveraging models that are common in control theory, such as dynamical systems, we can better exploit the interconnection between the input and output of a given system, which is often leveraged in CPS attacks. This has been addressed in the work of Urbina et al. (2016) and is a valuable tool in defending against attacks such as in drone spoofing.

The first consideration of control theorists has been, as is to be expected in problems of security, to develop methods and algorithms that achieve intrusion detection of malicious signals to the different components of CPS. Naturally, control theorists have focused mostly on injections to the physical interfaces of CPS, i.e., on attacks to the actuators and sensors of such systems. Depending on the model that the authors employed, we can broadly separate the literature into research conducted on continuous-time, discrete-time, and hybrid systems. This differentiation is important from an abstraction point of view since different scenarios can be more readily considered via each of these models.

Vamvoudakis et al. (2014) focused on the estimation problem of a binary variable in a network of sensors under Byzantine attacks, with no consideration of the system dynamics. After introducing evolution modeled via continuous-time systems, Pasqualetti et al. (2013) provide a framework to characterize various types of attacks for linear systems in the absence of uncertainties. Then they derive conditions for the detection and identification of these attacks through an observer-based analysis. A detector of replay attacks is developed in Hoehn and Zhang (2016), where the authors propose an excitation of their system in non-regular time intervals to reveal a naive attacker who has no knowledge of the system model. Detection for hybrid systems was also considered in Phillips et al. (2017), where tools for detecting a class of attacks affecting both the flow and the jump dynamics were derived. Temporal logic (Jones et al., 2014) has also been used with the goal of detecting adversarial attacks on CPS. Similar works have been conducted for discrete-time systems. Miao et al. (2016) use coding techniques to transform the system output and leverage the encoded information to increase the probability of detecting sensor and actuator attacks. In the same context of stochastic systems, the work in Guan and Ge (2017) provides a decentralized mechanism able to detect false data injection attacks when the sensors are jammed. For deterministic discrete-time systems with no uncertainties, Chen et al. (2016) characterize detectable and undetectable attacks. They derive a detector that manages to identify the detectable attacks without full knowledge of the initial condition of the system. The work of Fawzi et al. (2014) showed that if the attacker can compromise less than half of the sensors, then it is always possible to recover the state information. To relax this assumption, several switching-based schemes have been developed so that, rather than guaranteeing robustness of estimation or operation under attack, they opt to identify the attacked components and take them offline. Following this line of research, an attack detection filter and a passivity-based switching mechanism were introduced in An and Yang (2017); Yan et al. (2017)

Moving beyond simple sensors and actuators, control theorists have also considered robustness and security problems arising from the inclusion of data-driven learning algorithms in modern autonomous CPS, as well as in the underlying data transfer errors stemming from imperfect signal quantification and timing inconsistencies among physical components with separate local clocks. Specifically, Jiang and Jiang (2014); Modares et al. (2015) analyzed the performance of an off-policy reinforcement learning algorithm in the presence of errors created due to neural network approximation. It was shown that if the number of neurons employed to approximate the value function and the control policy is sufficiently large, then the off-policy algorithm still attains convergence. In Gao et al. (2022, 2016) an output-based off-policy learning-based optimal control approach was designed, which yields policies robust to dynamic uncertainties and denial-of-service attacks. Recently, the robustness of off-policy learning-based optimal control with respect to noise was revisited in Pang et al. (2021). In the context of clock offsets and quantization errors, Fridman and Dambrine (2009) study the input-to-state stability properties of a linear system in the presence of quantization errors, delays, and saturation via the use of Lyapunov–Krasovskii functionals. The problem of timing discrepancies between actuators and sensors is investigated in Wakaiki et al. (2017), where the discrepancies are modeled as parametric uncertainties. Expanding on this research, Okano et al. (2017) consider both clock mismatches and quantization errors. Finally, in Wakaiki et al. (2019) an adversarial scenario is analyzed, in which the effect of denial-of-service attacks on the system are considered in tandem with output quantization.

Most of the aforementioned approaches have been reactive in nature. Through our previous discussions on the relevant literature, it has become obvious that various fields of research have thrived in their pursuit of approaches to shield systems against attackers or in the development of detection mechanisms. These approaches, however, fail to take into account zero-day attacks to systems and disregard the dynamical nature of the abilities of cyber-attackers. This creates an asymmetric situation, in which the defenders employ static tools that are rigid in their adaptability, whereas the attackers can plan and perform extensive reconnaissance over lengthy periods of time while launching the actual attack in short bursts. MTD (Jajodia et al., 2011) is a defense paradigm that aims to minimize the inherent advantage the attacker has over the defender. Most CPS operate statically with respect to their structure, goals, and constraints. Such vulnerabilities offer to a persistent attacker the necessary time to exploit the system and develop appropriate strategies. MTD protocols aim to tackle this asymmetry by developing mechanisms that continually and unpredictably change the parameters of the system. Such unpredictability has three goals: to increase the cost of attacking, to limit the exposure of vulnerable components, and to deceive the opponent. Note that these proactive approaches were initially introduced in the computer science community, with applications to computer networks (Casola et al., 2014; Jafarian et al., 2012; Jajodia et al., 2012). Dunlop et al. (2011) apply the principles of MTD to constantly rotating Internet Protocol version 6 (IPv6) addresses. Similarly, in Lu et al. (2016) a related proactive defense strategy was formulated to deceive an attacker targeting nodes in a wireless network. Returning to our control-theoretic viewpoint, we strive to maintain the vision of a reactive, unpredictability-based defense scheme using a more formalized approach to MTD that was introduced in Zhuang et al. (2014) and led to an MTD entropy hypothesis framework that is more generally applicable. An MTD approach, closely related to the methods presented in this manuscript, was used in Weerakkody and Sinopoli (2015) to enlarge the dimension of the state space for the purpose of attack detection, rather than proactive defense based on an unpredictability measure.

1.3. The viewpoint of game theorists

Upon examining our endeavors in this monograph, the reader will notice that we approach a rather difficult problem in security, adversarial modeling. Not unexpectedly, there are various ad hoc methods of designing prediction models for the behavior of a malicious agent, the validity of which mostly lies upon very strict assumptions. To better structure our study, we use game theory as an extremely potent tool of interaction modeling while simultaneously exploring more modern and, perhaps, more realistic methods of deriving predictions out of a game theoretic framework. Even though game theory is an extremely active field of research, spanning multiple communities ranging from physicists to computer scientists and engineers, there have been concerns related to the difficulty arising in predicting the strategies and capabilities of potential adversaries. Well-known prediction models follow the concept of Nash equilibrium (Basar and Olsder, 1999), wherein all the agents involved are assumed to share the same decision-making mechanism and are informed participants in a game. However, in many games where human players are involved, equilibrium models fail to predict or explain observed behavior. An alternative explanation is that agents are boundedly rational, i.e., agents may make mistakes during the game. Decision makers may never play the exactly same game very often, but they may also extrapolate between games and learn from experiences. Several recent experimental studies (Crawford and Iriberri, 2007) suggest that decision makers' initial responses to games often deviate systematically from equilibrium and that structural non-equilibrium models (e.g., cognitive hierarchy) often out-predict equilibrium. Thus non-equilibrium models need to allow for players whose adjustment rules are not the best response to the adjustment rules of the others.

One of the first works on non-equilibrium game theoretic behavior in static environments has been reported in Brams and Kilgour (1988); Fudenberg et al. (1998); Fudenberg and Levine (1998); Tambe (2011). In such works, the authors state that for most purposes, the right models involve neither full rationality nor the extreme naivete of most stimulus-response models; "better" models have agents consciously but perhaps imperfectly trying to get a good payoff. Besides specifying the players' forecast rules, an analysis of learning must also address the issue of whether players try to influence their opponents' play. The work of Erev and Roth (1998); Roth and Erev (1995) examined reinforcement learning approaches for experimental games with unique equilibrium and discussed implications for developing a low-rationality framework, namely behavioral game theory. He et al. (2016) overcame the irrationality issue with a Win-or-Learn Fast in a mini-max-Q learning framework, but with the penalty of losing convergence assurance. Quantal response models (McKelvey and Palfrey, 1995) assume that the players' actions are perturbed equilibrium policies, altered by stochastic mistakes. Another class of non-equilibrium models depends on the computation of a finite number of best response strategies up to a certain level. Level-k thinking assumes that each player believes that all his opponents operate at the $(k-1)$th level of intelligence. On the other hand, in cognitive hierarchy models, each player believes that his opponents' levels follow a Poisson distribution. Both models are examined in Chong et al. (2016). These results have intrigued the CPS security community. For instance, Li et al. (2016) employ a cognitive hierarchy approach to train autonomous vehicles in real-world situations. Cognitive hierarchy was also used for cyber-physical security in Abuzainab et al. (2016) for the problem of distributed uplink random access for the Internet of Things.

CHAPTER TWO

Control-theoretic intrusion detection

2.1. Bellman-based detection

Throughout this book, we will investigate the problem of securing CPS via the viewpoint of control theory. As such, our initial concern will be to introduce the mathematical model that captures the interconnections of several components of a real system and their evolution over time. Although specific applications call for different approaches, our typical tool for such a description will be a state-space model of a dynamical systems, each endowed with input/output interfaces, actuators and sensors.

2.1.1 System and attack signal model

Let us consider the following set of differential equations, coupled with a set of algebraic ones:

$$\begin{aligned}\dot{x}(t) &= Ax(t) + Bu_a(t), \ \ t \geq 0, \\ y(t) &= C_a(t)x(t). \end{aligned} \tag{2.1}$$

Taking advantage of the modeling potency of differential equations, we denote by $x(t) \in \mathbb{R}^n$ an internal state of variables that completely and uniquely define the evolution of the system, by $u_a(t) \in \mathbb{R}^m$ the input signal to the system, which might be compromised by an adversary, by $y(t) \in \mathbb{R}^p$ the measured output, available to the designer and operator of the system, as well as the structural elements of our description, $A \in \mathbb{R}^{n \times n}$ being the plant matrix, $B \in \mathbb{R}^{n \times m}$ the input matrix, and $C_a(t) \in \mathbb{R}^{p \times n}$ the potentially compromised output matrix.

Following the well-known control theoretic methods, we imbue this system with LQR feedback controllers given as

$$u(x) = -\frac{1}{2}R^{-1}B^{\mathrm{T}}\nabla V(x), \ \ x \in \mathbb{R}^n,$$

where the optimal value function is given by the quadratic expression $V^\star(x) = x^{\mathrm{T}}Px$, $P \succ 0$, determined by a matrix P solving the Riccati equation

$$A^{\mathrm{T}}P + PA - PBR^{-1}B^{\mathrm{T}}P + Q = 0.$$

Control and Game Theoretic Methods for Cyber-Physical Security
https://doi.org/10.1016/B978-0-44-315408-9.00008-7

2.1.2 Bellman detection against actuator attacks

In this section, an intrusion detection mechanism is designed to identify the potentially corrupted controller. The attack detection signal will rely on the optimality property and on data measured along the – possibly corrupted – trajectories of the system. Based on a sampling mechanism, we denote the measurements of the state at the sampling instances as $x_c(t)$ and define the functions $\hat{V}(\cdot) := x_c^{\mathrm{T}} P x_c$. Intuitively, we obtain a sampled version of the optimal value function along the system real trajectories.

Lemma 2.1. *The error between the optimal and real (potentially attacked) trajectories under the integrable attack signal $\rho(t)$ over a closed time interval $[t_0, t_1]$ is bounded as*

$$\|e_x(t)\| \leq \alpha(t, \rho)\|x(t_0)\|,$$

where

$$\alpha(t, \rho) = \int_{t_0}^{t} \delta(\tau)\|I - \rho(t)\| e^{\int_{t_0}^{t} \delta(\sigma)\|I-\rho(\sigma)\| d\sigma} d\tau,$$

and

$$\delta(\tau) = \|e^{(A-BK)(\tau - t_0)}\| \|B\| \|K\|$$

with $K = R^{-1}B^T P$.

Proof. For ease of exposition, we denote the time interval under consideration as $[t_0, t_1]$. Let $x^\star(t)$ be the trajectory of the system driven by (3.7) under the absence of attacks, and let $x_c(t)$ be the actual trajectory. Also, we assume, without loss of generality, that for all $t \leq t_0$, there is no attack on the system, and therefore the optimal trajectory of the system coincides with the actual trajectory $x^\star(t_0) = x_c(t_0) = x(t_0)$.

The optimal and actual trajectories evolve according to

$$\dot{x}^\star(t) = (A - BK)x^\star(t), \ x^\star(t_0) = x(t_0),$$
$$\dot{x}_c(t) = (A - B\rho(t)K)x_c(t), \ x_c(t_0) = x(t_0).$$

Since we take into account attack signals that may be integrable but discontinuous, the trajectories $x_c(t)$ are defined in the sense of Carathéodory.

First, we consider the actual trajectory of the system $x_c(t)$, $t \in [t_0, t_1]$, according to

$$\dot{x}_c = (A - B\rho K)x_c \ , \ x_c(t_0) = x(t_0),$$

$$\dot{x}_c = (A - BK)x_c + (B(I - \rho)K)x_c. \tag{2.2}$$

The solution to (2.2), with $(B(I - \rho)K)x_c$ taken as a forcing term, is

$$x_c(t) = e^{(A-BK)(t-t_0)}x_c(t_0) + \int_{t_0}^{t} e^{(A-BK)(t-\tau)}B(I - \rho(\tau))Kx_c(\tau)\mathrm{d}\tau.$$

Taking norms yields

$$\begin{aligned}\|x_c(t)\| \leq \|e^{(A-BK)(t-t_0)}\|\,\|x(t_0)\| \\ + \int_{t_0}^{t} \|e^{(A-BK)(t-\tau)}\|\,\|B\|\,\|(I - \rho(\tau))\|\,\|K\|)\|x_c(\tau)\|\mathrm{d}\tau.\end{aligned}$$

Since each controller K renders the system stable, we know that the transition matrix of the closed-loop system will be upper bounded. Therefore we can introduce $\gamma = \max_t \|e^{(A-BK)(t-t_0)}\|$. Denoting $\delta(\tau) = \|e^{(A-BK)(\tau-t_0)}\|\,\|B\|\,\|K\|$, we have

$$\|x_c(t)\| \leq \gamma\|x_0\| + \int_{t_0}^{t} \delta(\tau)\|I - \rho(\tau)\|\,\|x_c(\tau)\|\mathrm{d}\tau.$$

It has been shown in Ames and Pachpatte (1997) that under assumptions of integrability, Gronwall-type inequalities hold for discontinuous functions inside the integral, such as $\delta(\tau)\|I - \rho(\tau)\|$. Applying these results yields a bound on the norm of the actual trajectory:

$$\|x_c(t)\| \leq \gamma\|x(t_0)\|e^{\int_{t_0}^{t} \delta(\tau)\|I-\rho(\tau)\|\mathrm{d}\tau}. \tag{2.3}$$

We can now define the error between the actual and optimal trajectories as

$$e_x(t) = x_c(t) - x^\star(t) \tag{2.4}$$

with dynamics given by

$$\begin{aligned}\dot{e}_x = \dot{x}_c - \dot{x}^\star = (A - B\rho K)x_c - (A - BK)x^\star \Rightarrow \\ \dot{e}_x = (A - BK)e_x + B(\rho - I)Kx_c.\end{aligned}$$

Due to the assumption that $x^\star(t_0) = x_c(t_0)$, we have $e(t_0) = 0$ and

$$e_x(t) = \int_{t_0}^{t} e^{(A-BK)(t-\tau)}B(\rho(\tau) - I)Kx_c(\tau)\mathrm{d}\tau,$$

which, after taking norms, yields

$$\|e_x(t)\| \leq \int_{t_0}^{t} \|e^{(A-BK)(t-\tau)}\| \|B\| \|I - \rho(\tau)\| \|K\| \|x_c(\tau)\| \mathrm{d}\tau.$$

Utilizing now the bound (2.3), we can further write

$$\begin{aligned}\|e_x(t)\| &\leq \int_{t_0}^{t} \|B\| \|I - \rho(\tau)\| \|K\| \|e^{(A-BK)(t-\tau)}\| \|x(t_0)\| e^{\int_{t_0}^{\tau} \delta(\sigma)\|I-\rho(\sigma)\|\mathrm{d}\sigma} \mathrm{d}\tau \\ &= \int_{t_0}^{t} \delta(\tau) \|I - \rho(\tau)\| \|x(t_0)\| e^{\int_{t_0}^{\tau} \delta(\sigma)\|I-\rho(\sigma)\|\mathrm{d}\sigma} \mathrm{d}\tau.\end{aligned}$$

By using

$$\alpha(t, \rho) = \int_{t_0}^{t} \delta(t) \|I - \rho(t)\| e^{\int_{t_0}^{t} \delta(\sigma)\|I-\rho(\sigma)\|\mathrm{d}\sigma} \mathrm{d}\tau,$$

for which $\alpha(t, \rho) \neq 0$ for all $\rho \neq I$, we can write the bound on the trajectory error as

$$\|e_x(t)\| \leq \alpha(\tau, \rho) \|x(t_0)\|,$$

thus completing the proof. □

Remark 2.1. We can see that $\alpha(\tau, \rho) = 0$ if and only if $\rho(t) = I$ for all $t \in [t_0, t_1]$. ■

Theorem 2.1. *Consider the system operating based on (3.7) and (3.8). Define the detection signal over a predefined time window $T > 0$ as*

$$e(t) = \hat{V}(x_c(t - T)) - \hat{V}(x_c(t)) - \int_{t-T}^{t} (x_c^T Q x_c + u^{\star T} R u^{\star}) d\tau. \tag{2.5}$$

Then the system is under attack if and only if $e(t) \neq 0$. The optimality loss due to the attacks, quantified by $\|e(t)\|$, is bounded for any injected integrable signal $\rho(t)$.

Proof. As was proven in Vrabie et al. (2013), Eq. (2.5) is the integral form of the Bellman equation. For the sampled value of the state at $t_1 = t - T$, we have that

$$\hat{V}(t - T) = x^{\mathrm{T}}(t - T)Px(t - T) = \min_u \left\{ \int_{t-T}^{t} (x^{\mathrm{T}} Q x + u^{\mathrm{T}} R u) \mathrm{d}\tau + \hat{V}(t) \right\}.$$

Since $P \succ 0$, we have

$$\hat{V}(t - T) = x^{\mathrm{T}}(t - T)Px(t - T)$$

$$= \min_u \left\{ \int_{t-T}^{t} (x^{\mathrm{T}}Qx + u^{\mathrm{T}}Ru)\mathrm{d}\tau \right\} + x^{\mathrm{T}}(t)Px(t).$$

For the accumulated cost utilizing the optimal input and the cost utilizing an arbitrary input u_a, we have

$$\begin{aligned}
\int_{t-T}^{t} (x^{\mathrm{T}}Qx + u^{\star}Ru^{\star})\mathrm{d}\tau &= \min_u \left\{ \int_{t-T}^{t} (x^{\mathrm{T}}Qx + u^{\mathrm{T}}Ru)\mathrm{d}\tau \right\} \\
&\leq \int_{t-T}^{t} (x^{\mathrm{T}}Qx + u_a^{\mathrm{T}}Ru_a)\mathrm{d}\tau \Rightarrow \\
\int_{t-T}^{t} (x^{\mathrm{T}}Qx + u^{\star}Ru^{\star})\mathrm{d}\tau &= \int_{t-T}^{t} (x^{\mathrm{T}}Qx + u_a^{\mathrm{T}}Ru_a)\mathrm{d}\tau - I(\rho).
\end{aligned}$$

Due to Assumption 3.4, the solution is unique. By extension the optimal cost over any time interval is also unique. Consequently, the system is attack-free when $I(\rho) = 0$.

For the boundedness part of the proof, we adopt the notation of Lemma 2.1. Along the actual trajectory of the system within a time interval $[t_0, t_1]$, the intrusion detection signal is

$$e(t) = x_c^{\mathrm{T}}(t_0)Px_c(t_0) - x_c^{\mathrm{T}}(t_1)Px_c(t_1) - \int_{t_0}^{t_1} (x_c^{\mathrm{T}}(\tau)Qx_c(\tau) + u^{\star\mathrm{T}}(x_c)Ru^{\star}(x_c))\mathrm{d}\tau.$$

Since the control signal utilized by the controller will be optimal, for all $t \geq 0$, we can write

$$\begin{aligned}
e(t) &= x_c^{\mathrm{T}}(t_0)Px_c(t_0) - x_c^{\mathrm{T}}(t_1)Px_c(t_1) \\
&\quad - \int_{t_0}^{t_1} \left(x_c^{\mathrm{T}}(\tau)Qx_c(\tau) + \left(-R^{-1}B^{\mathrm{T}}Px_c(\tau)\right)^{\mathrm{T}} R\left(-R^{-1}B^{\mathrm{T}}Px_c(\tau)\right)\right)\mathrm{d}\tau \Rightarrow \\
e(t) &= x_c(t_0)^{\mathrm{T}}Px_c(t_0) - x_c(t_1)^{\mathrm{T}}Px_c(t_1) - \int_{t_0}^{t_1} x_c^{\mathrm{T}}(\tau)\tilde{Q}x_c(\tau)\mathrm{d}\tau,
\end{aligned}$$

where $\tilde{Q} = Q + PBR^{-1}B^{\mathrm{T}}P \succ 0$. The positive definiteness is derived by the asymptotic stability property of the optimal closed-loop system. We substitute the actual trajectory $x_c(t)$, utilizing the trajectory error (2.4), to write

$$\begin{aligned}
e(t) &= x_c^{\mathrm{T}}(t_0)Px_c(t_0) - \left(x^{\star}(t_1) + e_x(t_1)\right)^{\mathrm{T}} P\left(x^{\star}(t_1) + e_x(t_1)\right) \\
&\quad - \int_{t_0}^{t_1} \left(x^{\star}(\tau) + e_x(\tau)\right)^{\mathrm{T}} \tilde{Q}\left(x^{\star}(\tau) + e_x(\tau)\right)\mathrm{d}\tau,
\end{aligned}$$

which can be rewritten as

$$e(t) = x_c^{\mathrm{T}}(t_0)Px_c(t_0) - x^{\star\mathrm{T}}(t_1)Px^\star(t_1) - \int_{t_0}^{t_1} x^\star(\tau)^{\mathrm{T}}\tilde{Q}x^\star(\tau)$$
$$- \Big\{ e_x^{\mathrm{T}}(t_1)Px^\star(t_1) + x^{\star\mathrm{T}}(t_1)Pe_x(t_1) + e_x^{\mathrm{T}}(t_1)Pe_x(t_1)$$
$$+ \int_{t_0}^{t} (e_x^{\mathrm{T}}(\tau)\tilde{Q}x^\star(\tau) + x^{\star\mathrm{T}}(\tau)\tilde{Q}e_x(\tau) + e_x^{\mathrm{T}}(\tau)\tilde{Q}e_x(\tau))\mathrm{d}\tau \Big\}.$$

The first three terms of this expression satisfy the integral form of the HJB equation. As a result, the residual terms that quantify the optimality loss due to the attack are

$$e(t) = -e_x^{\mathrm{T}}(t_1)Px^\star(t_1) + x^{\star\mathrm{T}}(t_1)Pe_x(t_1) + e_x^{\mathrm{T}}(t_1)Pe_x(t_1)$$
$$+ \int_{t_0}^{t} (e_x^{\mathrm{T}}(\tau)\tilde{Q}x^\star(\tau) + x^{\star\mathrm{T}}(\tau)\tilde{Q}e_x(\tau) + e_x^{\mathrm{T}}(\tau)\tilde{Q}e_x(\tau))\mathrm{d}\tau.$$

Taking norms and utilizing the fact that $x^\star(t) = e^{(A-BK)(t-t_0)}x(t_0)$ and Lemma 2.1, we can bound the norm of the intrusion detection signal as

$$\|e(t)\| \leq b(t,\rho)\|x(t_0)\|^2,$$

where

$$b(t,\rho) = 2\alpha(t,\rho)\|P\|\gamma + \alpha^2(t,\rho)\|P\|$$
$$+ \int_{t_0}^{t} \big(2\alpha(\tau,\rho)\|\tilde{Q}\|\gamma + \alpha^2(\tau,\rho)\|\tilde{Q}\|\big)\mathrm{d}\tau,$$

with the property that $b(t,\rho) = 0$ for all $t \in [t_0, t_1]$ if and only if $\rho(t) = I$. □

It is possible to extend the results of the previous section to consider noise in the actuation mechanism:

$$u_a(t) = \rho(t)u^\star(t) + w(t),$$

where $w(t)$ is a bounded but unknown disturbance with $\|w(t)\| \leq \bar{w}$.

Theorem 2.2. *System* (2.1), *equipped with the detection mechanism as defined in Theorem* 2.1, *under the effect of a disturbance* $w(t)$, *is compromised if*

$$\|e(t)\| \geq e_{thres}(t),$$

where e_{thres} are the dynamic thresholds of the form

$$e_{thres}(t) = 2\|\bar{w}\| \int_{t-T}^{t} \|Ru^{\star}(\tau)\| d\tau + \bar{\lambda}(R)\|\bar{w}\|^2.$$

Proof. First, we will consider the system in the absence of attacks and formulate the intrusion detection signal based on the data collected along the trajectories of the system. In other words, we can write

$$\begin{aligned}
e(t) &= \hat{V}(t-T) - \hat{V}(t) - \int_{t-T}^{t} (x^{\mathrm{T}} Qx + u_a^{\mathrm{T}} Ru_a) \mathrm{d}\tau \\
&= \hat{V}(t-T) - \hat{V}(t) - \int_{t-T}^{t} \big(x^{\mathrm{T}} Qx + (u^{\star} + w)^{\mathrm{T}} R(u^{\star} + w)\big) \mathrm{d}\tau \\
&= \hat{V}(t-T) - \hat{V}(t) - \int_{t-T}^{t} (x^{\mathrm{T}} Qx + u^{\star\mathrm{T}} Ru^{\star}) \mathrm{d}\tau \\
&\quad - \int_{t-T}^{t} (w^{\mathrm{T}} Ru^{\star} + u^{\star\mathrm{T}} Rw + w^{\mathrm{T}} Rw) \mathrm{d}\tau.
\end{aligned}$$

Leveraging the integral Bellman equality and taking norms yield

$$\begin{aligned}
&\|e(t)\| \le 2 \int_{t-T}^{t} \|w^{\mathrm{T}} Ru^{\star}\| \mathrm{d}\tau + \int_{t-T}^{t} \|w^{\mathrm{T}} Rw\| \mathrm{d}\tau \Rightarrow \\
&\|e(t)\| \le 2\|\bar{w}\| \int_{t-T}^{t} \|Ru^{\star}(\tau)\| \mathrm{d}\tau + T\bar{\lambda}(R)\|\bar{w}\|^2,
\end{aligned}$$

which is the adaptive threshold for the active controller i. □

Remark 2.2. Note that the adaptive threshold can be computed online utilizing only knowledge of the optimal input signal that the controller sends to the system (and not the potentially corrupted one). ■

Remark 2.3. In noisy environments, the system may be under attack, and the integral Bellman error may not cross the adaptive threshold. Thus the attack will remain undetected. However, attacks that have so little effect on the system become indistinguishable from random noise and do not degrade the performance of the system in a significant way. ■

2.1.3 Bellman detection against sensor attacks

We now extend the results of the previous subsection to show how the methods developed can be applied to the problem of observer design. We still consider that the system controller operates under full state feedback, and we focus on the observer problem separately, employing the results from Na et al. (2017).

The observer of (2.1) will be designed as a dynamic system sharing the same structural properties,

$$\dot{\hat{x}} = A\hat{x} + Bu + B\bar{u},$$
$$\hat{y} = C\hat{x}, \tag{2.6}$$

where $\hat{x}$ and $\hat{y}$ are the estimates of the state and output, respectively, and $\bar{u}$ denotes a "fictional" input, i.e., a correction term that forces the observer to track the actual system.

Following Durbha and Balakrishnan (2005); Na et al. (2017), to design the optimal $\bar{u}$, we define the optimization problem based on the following cost function for $t \geq 0$,

$$U^{\star}(\hat{x}) = \min_{\bar{u}} \int_t^{\infty} \left[(\hat{y} - y)^{\mathrm{T}} Q(\hat{y} - y) + \bar{u}^{\mathrm{T}} R\bar{u}\right] \mathrm{d}\tau.$$

Defining the Hamiltonian of the system as

$$H(\hat{x}, \bar{u}^{\star}, U^{\star}) = (\hat{y} - y)^{\mathrm{T}} Q(\hat{y} - y) + \bar{u}^{\star\mathrm{T}} R\bar{u}^{\star} + \nabla U^{\star\mathrm{T}}(A\hat{x} + Bu + B\bar{u}^{\star}) = 0. \tag{2.7}$$

We can now find the optimal control from the stationarity conditions $\frac{\partial H(\hat{x}, \bar{u}^{\star}, U^{\star})}{\partial \bar{u}^{\star}} = 0$. This leads to

$$\bar{u}^{\star} = -R^{-1} B^{\mathrm{T}} \nabla U^{\star}(\hat{x}).$$

Due to the quadratic structure of the cost functional and the linear structure of the dynamic system, it is known (Na et al., 2017) that the value function is quadratic in $\hat{x}(t)$, i.e., $U^{\star}(\hat{x}) = \hat{x}^{\mathrm{T}} G\hat{x}$ for some matrix $G \succ 0$, which means that the optimal "input" is

$$\bar{u}^{\star} = -R^{-1} B^{\mathrm{T}} G\hat{x}. \tag{2.8}$$

We will now introduce a detection signal based on the online, possibly compromised, estimations of the state, which we will denote $\hat{x}_c(t)$. For that reason, we introduce the function $\hat{U}(t) = \hat{x}_c^{\mathrm{T}} G\hat{x}_c,\ G \succ 0$.

Theorem 2.3. *Consider system (2.1) designed based on (2.7) and (2.8). Define the detection signal over a predefined time window $T > 0$ as*

$$e^s(t) = \hat{U}(\hat{x}_c(t-T)) - \hat{U}(\hat{x}_c(t)) - \int_{t-T}^{t} \left((y-\hat{y})^T Q(y-\hat{y}) + \bar{u}^{\star T} R \bar{u}^{\star}\right) d\tau. \tag{2.9}$$

Then the system is under attack if and only if $e^s(t) \neq 0$. Moreover, the optimality loss due to attacks is bounded for any injected signal $\rho^s(t)$.

Proof. The first part of the proof follows from Theorem 3.1 for the optimal control problem formulated in this section and is based on the uniqueness of optimal solutions for a given initial condition. However, to compute the bound on the optimality/observation loss, we define the measurement error $\tilde{y} = y - \hat{y}$. Then the detection signal is

$$e^s(t) = \hat{U}(\hat{x}_c(t-T)) - \hat{U}(\hat{x}_c(t)) - \int_{t-T}^{t} (\tilde{y}^{\mathrm{T}} Q_i \tilde{y} + \bar{u}^{\mathrm{T}} R \bar{u}_i) \mathrm{d}\tau.$$

Note that, in the presence of attacks,

$$y - \hat{y} = \rho^s C\hat{x} - C\hat{x} = \tilde{y} + \delta^s(\rho^s),$$

where $\delta^s(\rho^s) = (I - \rho^s)C\hat{x}$. Therefore the detection signal under attack is

$$\begin{aligned} e^s(t) &= \hat{U}(\hat{x}_c(t-T)) - \hat{U}(\hat{x}_c(t)) \\ &\quad - \int_{t-T}^{t} \left((\tilde{y} + \delta^s(\rho^s))^{\mathrm{T}} Q(\tilde{y} + \delta^s(\rho^s)) + \bar{u}^{\mathrm{T}} R \bar{u}_i\right) \mathrm{d}\tau. \end{aligned}$$

Expanding the quadratic terms, the residual detection signal becomes

$$e^s(t) = -\int_{t-T}^{t} \left(\delta^{s\mathrm{T}}(\rho^s) QC\hat{x} + (C\hat{x})^{\mathrm{T}} Q\delta^s(\rho^s) + \delta^{s\mathrm{T}}(\rho^s) Q\delta^s(\rho^s)\right) \mathrm{d}\tau.$$

Using the Cauchy–Schwarz inequality, we can bound the norm of the error as

$$\|e^s(t)\| \leq 2\int_{t-T}^{t} \|C\hat{x}\| \|\delta^s(\rho^s)\| + T\bar{\lambda}(Q)\|\delta^s(\rho^s)\|,$$

which completes the proof. □

Remark 2.4. The bound on the optimality/observation loss can be quantified more easily, because the injected attack does not directly affect the dynamics in system (2.1), rather it behaves like a noise in the cost term defined by the output error $(y - \hat{y})^{\mathrm{T}} Q(y - \hat{y})$. ■

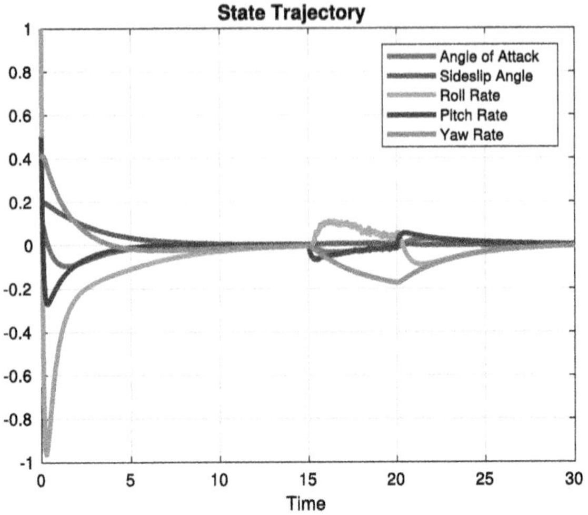

Figure 2.1 The evolution of the states in the presence of actuator attacks. The attack takes place for $t \in [15, 20]$.

2.1.4 Simulation results

To show the effectiveness of our approaches, we will use a linearized five-dimensional model of the ADMIRE benchmark aircraft (Yu and Jiang, 2012). The model has seven redundant actuators and two redundant sensors. In Fig. 2.1, we can see the evolution of the states under an attack signal for $t \in [15, 20]$, where the intrusion detection system was utilized. In Fig. 2.2, we see the evolution of the integral Bellman error. Although its magnitude is small, due to the absence of stochastic noise, the integral Bellman error is still able to detect the attack.

In Fig. 2.3, we consider a system under actuator attacks in the presence of system noise. Specifically, the noise has known upper bound $\|w\| = 0.5$, whereas the attack is a random signal with maximum value of 0.3. We note from Fig. 2.4 that the intrusion detection signal is able to detect the injected signal despite the noise.

Also, we consider the optimal state estimation problem for the ADMIRE aircraft utilizing the optimal observer framework. The state evolution of the observer is shown in Fig. 2.5. We notice that the attacker injects a relatively small bias to the estimated signal. For the sensor attacks, we take into account noise (with known statistics) on the measurements and show the evolution of the integral Bellman error and of the adaptive threshold in Fig. 2.6. Even though the discrepancy between the estimated angle of attack and the actual one is small relative to the measurement noise, the

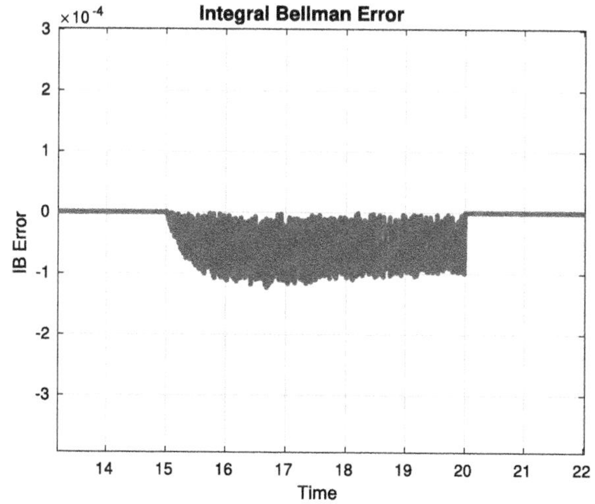

Figure 2.2 The evolution of the integral Bellman error. For the time interval where the attacker inputs an adversarial signal, $t \in [15, 20]$, the integral Bellman error is nonzero, which is sufficient to achieve intrusion detection in the absence of stochastic noise.

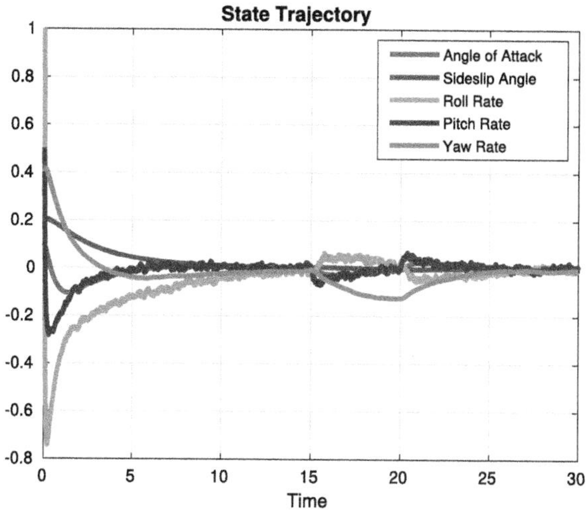

Figure 2.3 State evolution of a system with noise and actuator attacks. We note that the attack takes place at $t = [15, 20]$.

integral Bellman error manages to detect the attack. Specifically, during $t \in [14, 16]$, we detect the difference between the estimated error (due to sensor noise) and the one induced by the attack.

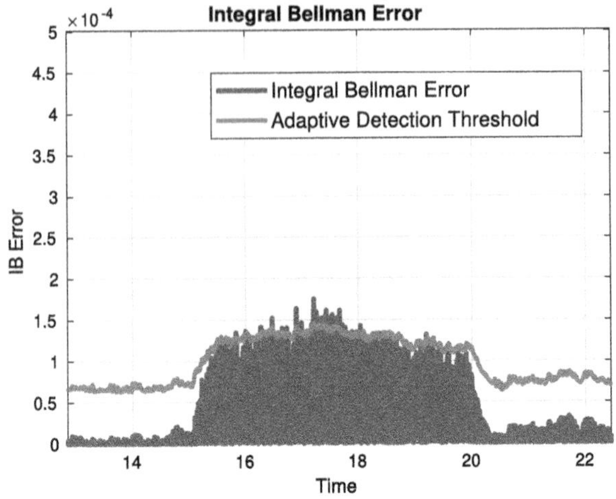

Figure 2.4 Evolution of the integral Bellman error for the system under attack and of the adaptive threshold that takes into account the system noise. Despite the magnitude of the attack being smaller than the noise, the proposed algorithm can detect the intrusion.

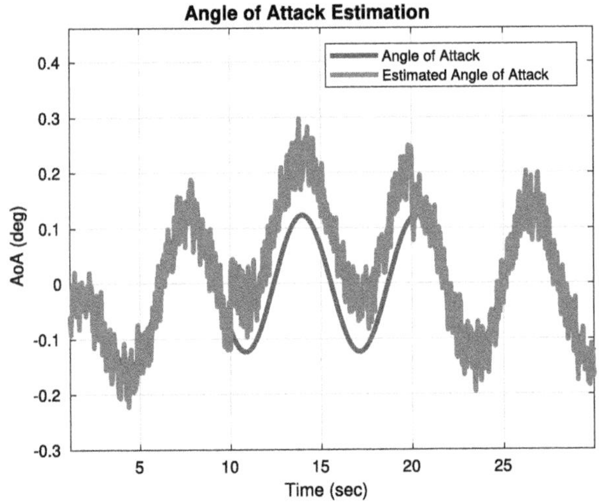

Figure 2.5 Optimal state estimation under noisy measurements and injected sensor attack. The attacker corrupts the output of the sensor for $t \in [10, 20]$ by adding a constant bias.

2.2. Statistics-based detection

In this section, we propose a statistics-intrusion detection mechanism to detect replay attacks with the help of watermarking signals. Replay

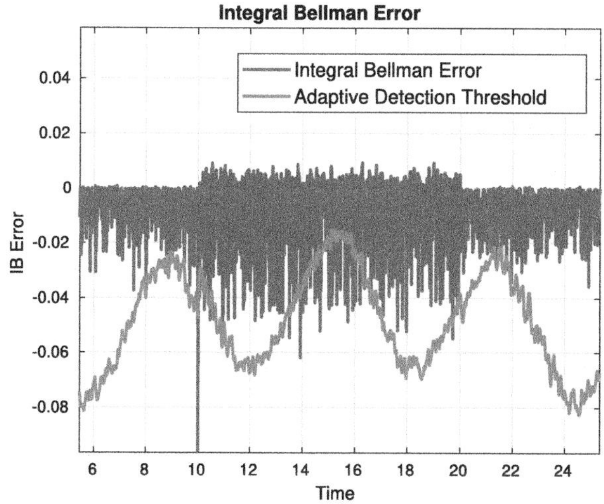

Figure 2.6 The evolution of the integral Bellman error and adaptive threshold for successful state reconstruction in the presence of sensor attacks. Even considering the noise of the sensors, the attack is detected.

attacks are common in CPS since adversaries do not need to have access to the system dynamics. The attack strategies of replay adversaries are recording a sequence of sensor measurements for a period of time and later modifying the current sensor measurement to the recorded signals. The key idea of the watermark-based detection mechanism is to add authentication watermark signals that follow a Gaussian distribution into optimal control signals. Utilizing hypothesis testing based on the Neyman–Pearson lemma, an alarm signal is computed at every instant of time. When the alarm signal is greater than a specified threshold, it implies that the system is under replay attacks. Specifically, we consider a discrete-time LTI system with added optimal watermarking signals. The designs of Neyman–Pearson detector and the optimal watermarking signals are derived. A data-based framework is proposed.

2.2.1 System model and replay attack strategy

Consider the following LTI discrete-time system for $k \in \mathbb{Z}$:

$$x_{k+1} = Ax_k + Bu_k + w_k, \tag{2.10}$$

$$y_k = Cx_k + v_k, \tag{2.11}$$

where $x_k \in \mathbb{R}^n$ is the state, $u_k \in \mathbb{R}^l$ is the control input, $y_k \in \mathbb{R}^m$ is the output, and $A \in \mathbb{R}^{n\times n}$, $B \in \mathbb{R}^{n\times l}$, and $C \in \mathbb{R}^{m\times n}$ are the state, control input, and output matrices, respectively. In this work, the process noise $w_k \in \mathbb{R}^n$ is assumed to be a zero-mean Gaussian noise with positive definite covariance $\Sigma_w \succ 0$, denoted as $w_k \sim \mathcal{N}(0, \Sigma_w)$. Similarly, the measurement noise $v_k \in \mathbb{R}^m$ is assumed to follow a zero-mean Gaussian distribution with positive definite covariance $\Sigma_v \succ 0$, denoted as $v_k \sim \mathcal{N}(0, \Sigma_v)$. In addition, the initial condition x_0 is assumed to follow a zero-mean Gaussian distribution with positive definite covariance $\Sigma_0 \succ 0$, denoted as $x_0 \sim \mathcal{N}(0, \Sigma_0)$. It is assumed that the signals w_k and v_k are uncorrelated and that the initial condition x_0 is independent of process and measurement noises.

Assumption 2.1. Given that CPS operate for an extended period of time, they are operating at a steady state. ■

Assumption 2.2. The pairs (A, B) and (A, C) are controllable and observable, respectively. ■

Information about full states is not always accessible for feedback. Consider the following observer for $k \in \mathbb{Z}$,

$$\hat{x}_{k+1} = A\hat{x}_k + Bu_k + Ke_k, \tag{2.12}$$

$$y_k = C\hat{x}_k + e_k, \tag{2.13}$$

where $\hat{x}_k \in \mathbb{R}^n$ denotes the estimated state, K stands for the steady-state Kalman filter gain, whereas the term $e_k \in \mathbb{R}^m$ is known as the innovation with covariance $E_k = \text{cov}(e_k) = \mathbf{E}(e_k e_k^{\mathrm{T}})$, where $\mathbf{E}(\cdot)$ denotes the expectation operator. The innovation E_k is calculated as $E_k = CSC^{\mathrm{T}} + \Sigma_v$. The steady-state Kalman gain is computed as $K = ASC^{\mathrm{T}}(CSC^{\mathrm{T}} + \Sigma_v)^{-1}$, where S stands for the covariance of the estimation error $\varepsilon_k := x_k - \hat{x}_k$. The covariance of the estimation error $\text{cov}(\varepsilon_k) = \mathbf{E}(\varepsilon_k \varepsilon_k^{\mathrm{T}})$ is calculated by solving the algebraic Riccati equation (ARE) $S = ASA^{\mathrm{T}} + \Sigma_w - ASC^{\mathrm{T}}\left(CSC^{\mathrm{T}} + \Sigma_v\right)^{-1} CSA^{\mathrm{T}}$. We can obtain the covariance of the estimated state $X_k = \text{cov}(\hat{x}_k) = \mathbf{E}(\hat{x}_k \hat{x}_k^{\mathrm{T}})$ by solving the following ARE for $k \in \mathbb{Z}$:

$$X_k = AX_kA^{\mathrm{T}} + ASC^{\mathrm{T}}\left(CSC^{\mathrm{T}} + \Sigma_v\right)^{-1} CSA^{\mathrm{T}}. \tag{2.14}$$

The optimal control law for $k \in \mathbb{Z}$ is

$$u_k^{\star} = L\hat{x}_k \ \forall \hat{x}_k \tag{2.15}$$

with the feedback gain $L = -(B^{\mathrm{T}}PB + R/\gamma)^{-1}B^{\mathrm{T}}PA$, where P is found by solving the ARE $P = \gamma A^{\mathrm{T}}PA + C^{\mathrm{T}}QC - \gamma A^{\mathrm{T}}PB(B^{\mathrm{T}}PB + R/\gamma)^{-1}B^{\mathrm{T}}PA$.

Specifically speaking, the replay attack strategies are described as follows:

- *Step 1.* Adversaries record a sequence of sensory measurements from time instant k_1 to $k_1 + T$, where $T \in \mathbb{Z}^+$ is chosen by adversaries to be large enough to guarantee that the sequence can be replayed for an extended period of time.
- *Step 2.* Adversaries replace the current sensory measurement y_k with the recorded one, i.e., $y'_k = y_{k-\Delta k}$, from time instant k_2 to $k_2 + T$, where y'_k denotes replayed signals, and $\Delta k := k_2 - k_1$.

2.2.2 Watermarking-based detection mechanism

To actively defend against replay attacks, physical watermarking signals are injected into the control input to serve as an authentication. Thus the overall control input u_k of the system is the sum of the optimal control input and watermarking signals, $u_k = u_k^\star + \phi_k$. It is assumed that watermarking signals ϕ_k follow a zero-mean Gaussian distribution with positive definite covariance $U \succ 0$, i.e., $\phi_k \sim \mathcal{N}(0, U)$. By adding watermarking signals to the system the distributions of measurement output without replay attacks and under replay attacks will be different, and thus a detector can be designed due to such a statistical difference with the help of hypothesis testing.

The distributions of the output data, with and without considering replay attacks, are characterized as follows. Given the system dynamics (2.10) and (2.11), it can be shown that the output signals without replay attacks for $k \in \mathbb{Z}$ are

$$\begin{aligned} y_k &= \sum_{t=0}^{k-1} CA^t B\phi_{k-1-t} + \sum_{t=0}^{k-1} CA^t B u^\star_{k-1-t} + \sum_{t=0}^{k-1} CA^t w_{k-1-t} + v_k + CA^k x_0 \\ &= \underbrace{\sum_{t=0}^{k-1} CA^t B\phi_{k-1-t}}_{\text{first term}} + \underbrace{\sum_{t=0}^{k-1} CA^t BL\hat{x}_{k-1-t}}_{\text{second term}} + \underbrace{\sum_{t=0}^{k-1} CA^t w_{k-1-t} + v_k + CA^k x_0}_{\text{third term}} . \end{aligned}$$

For simplicity, we split y_k into three terms and denote them for $k \in \mathbb{Z}$ as

$$\Phi_k := \sum_{t=0}^{k} CA^t B\phi_{k-t}, \tag{2.16}$$

$$\varrho_k := \sum_{t=0}^{k} CA^t BL\hat{x}_{k-t}, \tag{2.17}$$

$$\theta_k := \sum_{t=0}^{k} CA^t w_{k-t} + v_{k+1} + CA^{k+1} x_0. \tag{2.18}$$

Thus the measurement output for $k \in \mathbb{Z}$ is characterized by

$$y_k = \Phi_{k-1} + \varrho_{k-1} + \theta_{k-1}. \tag{2.19}$$

From (2.16) we can see that Φ_{k-1} is a zero-mean Gaussian distribution with covariance given by

$$\Upsilon := \sum_{\tau=0}^{\infty} (CA^\tau B) U (CA^\tau B)^{\mathrm{T}}. \tag{2.20}$$

Similarly, we can observe from (2.17) that ϱ_{k-1} is a zero-mean Gaussian distribution with covariance given by

$$\Gamma := \sum_{\tau=0}^{\infty} [CA^\tau BL] X_k [CA^\tau BL]^{\mathrm{T}}. \tag{2.21}$$

By observing (2.18) we can see that θ_{k-1} is a zero-mean Gaussian distribution with covariance given by $\Theta := C\Sigma C^{\mathrm{T}} + \Sigma_v$, where Σ stands for the covariance of state x_k for the dynamical system without control input, i.e., $x_{k+1} = Ax_k + w_k$, $y_k = Cx_k + v_k$, and Σ satisfies $\Sigma = A\Sigma A^{\mathrm{T}} + \Sigma_w$.

In the case that the system is under replay attacks, the replayed output is described for $k \in \mathbb{Z}$ as

$$y'_k = y_{k-\Delta k} = \Phi_{k-1-\Delta k} + \varrho_{k-1-\Delta k} + \theta_{k-1-\Delta k}. \tag{2.22}$$

From (2.19) we can conclude that in the nominal (attack-free) case, given collected data $\phi_0, \phi_1, \ldots, \phi_{k-1}$ and $u_0, u_1, \ldots, u_{k-1}$, the measurement output y_k converges to a Gaussian distribution with mean $\Phi_{k-1} + \varrho_{k-1}$ and covariance Θ, denoted as $y_k \sim \mathcal{N}_0(\Phi_{k-1} + \varrho_{k-1}, \Theta)$. In the case that the system is under replay attacks, the parameter Δk is unknown to the system operator even though it is deterministic. Therefore from (2.22) we can conclude that the output y'_k converges to a zero-mean Gaussian distribution with covariance $\Upsilon + \Gamma + \Theta$, denoted as $y'_k \sim \mathcal{N}_1(0, \Upsilon + \Gamma + \Theta)$.

Since y_k and y'_k follow two different Gaussian distributions, we can design a detector to distinguish this statistical difference through hypothesis testing. We will now consider the following binary hypothesis testing with the null hypothesis given by $\mathcal{H}_0$ and the alternative hypothesis given by $\mathcal{H}_1$. Their definitions are given as follows:

- $\mathcal{H}_0$: The measurement output y_k follows the Gaussian distribution $\mathcal{N}_0(\Phi_{k-1} + \varrho_{k-1}, \Theta)$.
- $\mathcal{H}_1$: The measurement output y'_k follows the Gaussian distribution $\mathcal{N}_1(0, \Upsilon + \Gamma + \Theta)$.

Based on the Neyman–Pearson lemma (Scharf and Demeure, 1991), the alarm signal of the Neyman–Pearson detector for the hypothesis $\mathcal{H}_0$ versus the hypothesis $\mathcal{H}_1$ is calculated for $k \in \mathbb{Z}$ as

$$\begin{aligned} g_k = & (y_k - \Phi_{k-1} - \varrho_{k-1})^{\mathrm{T}} \Theta^{-1} (y_k - \Phi_{k-1} - \varrho_{k-1}) \\ & - y_k^{\mathrm{T}} (\Upsilon + \Gamma + \Theta)^{-1} y_k. \end{aligned} \tag{2.23}$$

The alarm signal g_k is then compared to a predetermined threshold ξ, which is tuned based on the detection performance, such as detection and false alarm rates. Note that $g_k \geq \xi$ implies that the hypothesis $\mathcal{H}_1$ is valid and thus the system is under replay attacks. Otherwise, the null hypothesis $\mathcal{H}_0$ holds with $g_k < \xi$, and the system is under normal operation. A data-based framework to determine the optimal control policies and to design the Neyman–Pearson detector by a data-driven method is developed in Zhai and Vamvoudakis (2021). Furthermore, based on the proposed scheme, a switching watermarking-based detection scheme that incorporates unpredictability into the watermarking generation is investigated in Zhai et al. (2021).

2.2.3 Simulation results

To showcase the effectiveness of the proposed watermarking-based detection scheme, consider a third-order F-16 autopilot aircraft plant. The discrete-time plant model of the F-16 aircraft is given by Stevens et al. (2015)

$$A = \begin{bmatrix} 0.906488 & 0.0816012 & -0.0005 \\ 0.0741349 & 0.90121 & -0.000708383 \\ 0 & 0 & 0.132655 \end{bmatrix},$$

$$B = \begin{bmatrix} -0.00150808 \\ -0.0096 \\ 0.867345 \end{bmatrix}, \quad D = \begin{bmatrix} 0.00951892 \\ 0.00038373 \\ 0 \end{bmatrix}.$$

The system states are $x = [\alpha \quad q \quad \delta_e]^{\mathrm{T}}$, where α is the angle of attack, q is the pitch rate, and δ_e is the elevator deflection angle. The control input u represents the elevator actuator voltage, and the disturbance input d represents the wind disturbance gusting on the angle of attack. The output measures states, i.e., $y_k = x_k$. It is assumed that the system is subjected to a zero-mean white noise with $\Sigma_w = 0.01\ \mathbf{I}_3$ and $\Sigma_v = 0.01\ \mathbf{I}_3$. The parameters $\gamma = 0.8$ and $\alpha = 1$ are selected in the simulation. The initial condition is selected as $x_0 = [4\ 2\ 5\ ]^{\mathrm{T}}$. Moreover, we assume that the attacker records output signals of the previous 400 seconds and replays from the 401th second.

The evolution of the kernel matrix $\bar{P}$ is described in Fig. 2.7, which shows the convergence of the learning process by ADP. Figs. 2.8 and 2.9 show the optimal control input u_k, the worst-case disturbances d_k, and the optimal watermarking signals ϕ_k, as well as the evolution of the system states and the output. We can see that with the input and watermarking signals designed by the proposed data-based approach, the system does not explode in the presence of replay attacks.

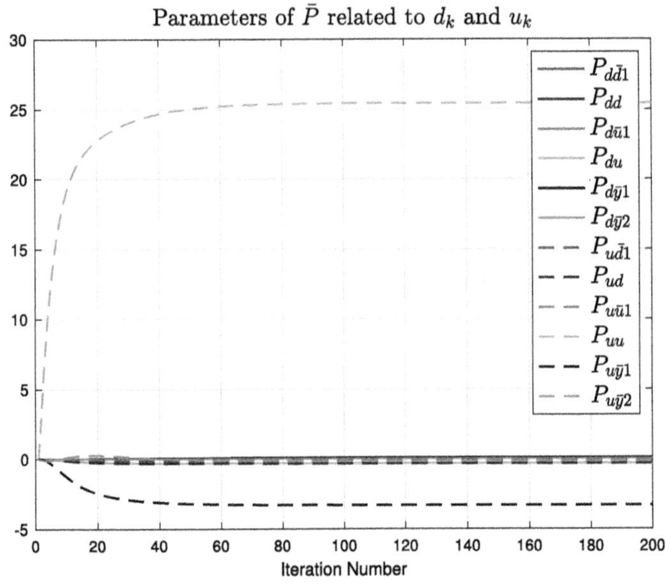

Figure 2.7 Evolution of parameters of $\bar{P}$ for the F-16 aircraft.

The comparison of alarm signals of the Neyman–Pearson detector between model-based and data-based approaches is shown in Fig. 2.10. We can see that the two signals overlap well, and a close agreement between the model-based and data-based approaches is achieved. After the 400th second, the alarm signals have a drastic jump implying the presence of replay attacks during this period. The model-based detector has some fluctuations at the beginning due to a lack of enough data during the initial phase.

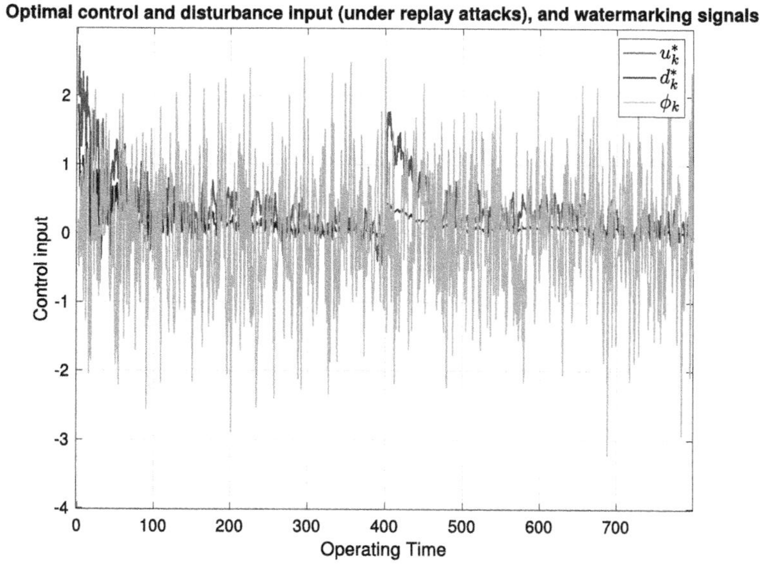

Figure 2.8 Evolution of system input and watermarking signals.

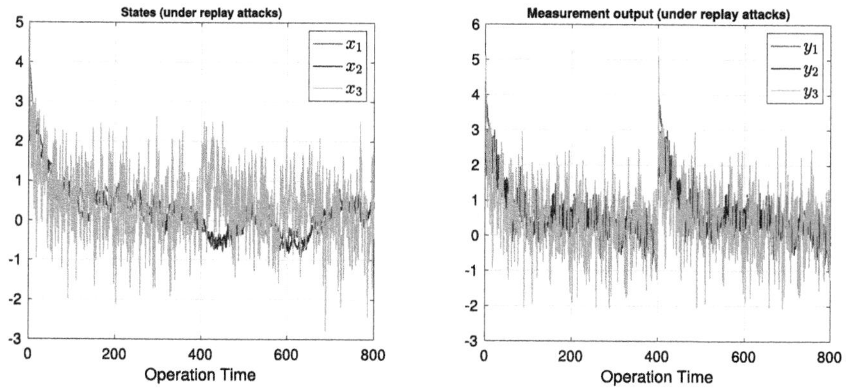

Figure 2.9 Evolution of system states and output.

2.3. Optimization-based detection

In this section, we consider the problem of detecting actuator attacks in continuous-time systems under the effect of unknown bounded deterministic disturbances. Unlike the previous sections, we will assume that the system state is available for measurement only at particular time instants, even if the system evolves in continuous time. This setup is especially com-

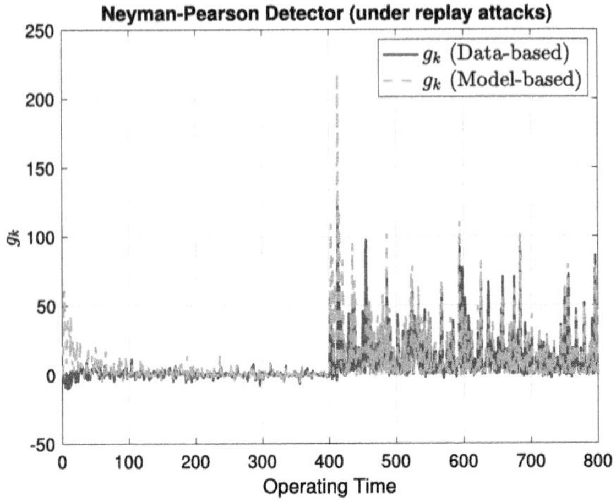

Figure 2.10 Comparison between alarm signals using the model-based and data-based approaches for the F-16 aircraft. Replay attacks record sensory output between the 0th and 400th second and then replay such recorded data between the 401th and 800th second.

mon in CPS, which comprise both discrete and continuous components due to the integration of software and physical systems.

2.3.1 Basic framework and assumptions

Consider the continuous-time dynamical system for $t \geq t_0 \geq 0$

$$\dot{x}(t) = Ax(t) + Bu(t) + Dd(t) + Ka(t), \;\; x(t_0) = x_0, \tag{2.24}$$

where $x: [t_0, \infty) \to \mathbb{R}^n$ is the state, $u: [t_0, \infty) \to \mathbb{R}^{m_u}$ is the control input of the operator, $a: [t_0, \infty) \to \mathbb{R}^{m_a}$ is an actuator attack, $d: [t_0, \infty) \to \mathbb{R}^{m_d}$ is an exogenous deterministic disturbance, and $A \in \mathbb{R}^{n\times n}$, $B \in \mathbb{R}^{n\times m_u}$, $D \in \mathbb{R}^{n\times m_d}$, and $K \in \mathbb{R}^{n\times m_a}$ are the system matrices. We consider the following assumptions.

Assumption 2.3. The deterministic exogenous disturbance d is unknown and bounded with respect to time, so that $\sup_{t\in[t_0, \infty)} \|d(t)\| < \Delta$ with known constant $\Delta > 0$. ■

Assumption 2.4. The state vector $x(t_k) = x_k$ is available for measurement only at specific time instants t_k, $k \in \mathbb{N}$, where $\xi_2 > t_{k+1} - t_k > \xi_1$ and $\xi_1, \xi_2 > 0$. ■

The operator's goal is, at every time instant t_k, $k \in \mathbb{N}_+$, to evaluate whether system (2.24) was under an actuator attack over $t \in [t_{k-1}, t_k]$ based on the measurements x_k and x_{k-1} of the state vector. Evidently, this goal can concern only attacks that are detectable in the sense that their impact on the system as captured by the state measurements x_k, $k \in \mathbb{N}$, is not justified by the uncertainty satisfying Assumption 2.3.

To characterize detectable actuator attacks more concretely, let us define an auxiliary model of system (2.24) in an attack-free scenario as

$$\dot{x}_f(t) = Ax_f(t) + Bu(t) + Dd_f(t), \quad t \geq t_0 \geq 0, \tag{2.25}$$

where $x_f : [t_0, \ \infty) \to \mathbb{R}^n$ is the state in the attack-free scenario, and $d_f : [t_0, \ \infty) \to \mathbb{R}^{m_d}$ is an uncertainty satisfying the norm bound defined in Assumption 2.3. Then, given two distinct state measurements x_{k-1} and x_k, $k \in \mathbb{N}_+$, of the nominal system (2.24) at the time instants t_{k-1} and t_k, respectively, we will declare that the system is under a detectable actuator attack during the interval $[t_{k-1}, \ t_k]$ if there exists an *admissible* disturbance over that interval. The following two definitions describe admissible disturbances and detectable attacks more clearly.

Definition 2.1. Let d_a be a mapping from $[t_{k-1}, \ t_k]$ to $\mathbb{R}^{m_d}$. Then such a mapping is called an *admissible disturbance* for system (2.24) over $[t_{k-1}, \ t_k]$, $k \in \mathbb{N}_+$, if:

1. given two measurements $x_k, \ x_{k-1} \in \mathbb{R}^n$ of $x(t)$ at t_k, t_{k-1}, we have that $x_f(t_k) = x_k$ if we let $x_f(t_{k-1}) = x_{k-1}$ and $d_f(t) = d_a(t)$ in (2.25) for all $t \in [t_{k-1}, t_k]$;
2. $\sup_{t\in[t_{k-1}, t_k]} \left\| d_a(t) \right\| < \Delta$. ■

Definition 2.2. System (2.24) is under a *detectable* actuator attack over the time interval $[t_{k-1}, \ t_k]$, $k \in \mathbb{N}_+$, if the set of admissible disturbances for (2.24) over this interval is empty. ■

Apparently, to evaluate the presence of detectable actuator attacks, it suffices to verify the nonexistence of an admissible disturbance for (2.24).

2.3.2 A sufficient and necessary condition for detection

We will now provide a mechanism to verify the presence of detectable attacks or, equivalently, the nonexistence of admissible disturbances, using only discrete samples of the system state. To this end, define the attack-free and disturbance-free version of (2.24) as

$$\dot{\bar{x}}(t) = A\bar{x}(t) + Bu(t), \ \ \bar{x}(t_{k-1}) = x_{k-1}, \ \ t \in [t_{k-1}, \ t_k], \ \ k \in \mathbb{N}_+,$$

where $\bar{x}:[t_{k-1},\ t_k]\to\mathbb{R}^n$. The dynamics of the difference $\tilde{x}_f(t):=x_f(t)-\bar{x}(t)$ is then given by

$$\dot{\tilde{x}}_f(t)=A\tilde{x}_f(t)+Dd_f(t),\quad \tilde{x}_f(t_{k-1})=0,\ t\in[t_{k-1},\ t_k],\ k\in\mathbb{N}_+. \tag{2.26}$$

The following fact, on which the design of our detection mechanism will be based, is a result of the definition of an admissible disturbance.

Fact 2.1. An admissible disturbance exists for system (2.24) over $t\in[t_{k-1},\ t_k]$, $k\in\mathbb{N}_+$, if and only if the terminal condition

$$\tilde{x}_f(t_k)=x_k-e^{A(t_k-t_{k-1})}x_{k-1}-\int_{t_{k-1}}^{t_k}e^{A(t_k-\tau)}Bu(\tau)\mathrm{d}\tau \tag{2.27}$$

holds for $\tilde{x}_f$ following the trajectories of (2.26) for some input d_f in (2.26) satisfying $\sup_{t\in[t_{k-1},t_k]}\left\|d_f(t)\right\|<\Delta$.

Remark 2.5. Fact 2.1 provides us with a reachability condition, which cannot be satisfied if system (2.24) is under a detectable attack. In essence, it states that a detectable attack is present if the terminal state (2.27) of system (2.26) cannot be reached by beginning from the origin at $t=t_{k-1}$ due to an uncertainty satisfying Assumption 2.3. ■

To check whether the feasibility condition of Fact 2.1 holds, we can define an infinite-dimensional optimization over the dynamics of (2.26) that has the conditions of Fact 2.1 as constraints. Then to verify the existence of a detectable attack, it suffices to check the feasibility of that optimization problem. More specifically, consider the minimization problem

$$\min_{d_f:[t_{k-1},\ t_k]\to\mathbb{R}^{m_d}}\quad \int_{t_{k-1}}^{t_k} r(\tilde{x}_f(\tau),d_f(\tau))\mathrm{d}\tau, \tag{2.28}$$

$$\text{s.t.}\quad \dot{\tilde{x}}_f(t)=A\tilde{x}_f(t)+Dd_f(t),\ \forall t\in[t_{k-1},\ t_k], \tag{2.29}$$

$$\tilde{x}_f(t_{k-1})=0, \tag{2.30}$$

$$\tilde{x}_f(t_k)=\tilde{x}_{fk}, \tag{2.31}$$

where we consider that

$$\begin{aligned}
\tilde{x}_{fk}&:=x_k-e^{A(t_k-t_{k-1})}x_{k-1}-\int_{t_{k-1}}^{t_k}e^{A(t_k-\tau)}Bu(\tau)\mathrm{d}\tau,\\
r(\tilde{x}_f(t),d_f(t))&:=\sum_{i=1}^{m_d}\int_0^{d_{fi}(t)}\phi^{-1}\left(\frac{w}{\Delta}\right)\mathrm{d}w,\\
\phi(w)&:=\frac{2}{\pi}\arctan(w),
\end{aligned} \tag{2.32}$$

with $d_f(t) = [d_{f1}(t) \ \ldots \ d_{fm_d}(t)]^{\mathrm{T}}$. Note that the integrand r of the cost to be optimized is a sigmoid-based convex non-quadratic functional inspired by Lyshevski (1998), which is bounded if and only if $\sup_{t\in[t_{k-1},t_k]} \|d_f(t)\| < \Delta$. Thus, using Fact 2.1, a feasible solution to the optimization problem (2.28) subject to (2.29)–(2.31) exists if and only if an admissible disturbance exists for system (2.24). Based on this logic, we have the following lemma, which provides a necessary and sufficient condition for the existence of a detectable attack.

Lemma 2.2. *System (2.24) is under a detectable attack over the interval $[t_{k-1},\ t_k]$, $k \in \mathbb{N}_+$, if and only if the equation*

$$\bar{x}_{fk} = -\int_{t_{k-1}}^{t_k} \Delta e^{A(t_k-\tau)} D\phi\left(D^{\mathrm{T}} e^{A^{\mathrm{T}}(t_k-\tau)} v\right) \mathrm{d}\tau \tag{2.33}$$

has an empty set of solutions with respect to the constant vector $v \in \mathbb{R}^n$.

Proof. Define the Hamiltonian of the optimization problem (2.28)–(2.31) as

$$H(\bar{x}_f(t), d_f(t), \lambda(t)) := \sum_{i=1}^{m_d} \int_0^{d_{fi}(t)} \phi^{-1}\left(\frac{w}{\Delta}\right) \mathrm{d}w + \lambda^{\mathrm{T}}(t)\left(A\bar{x}_f(t) + Dd_f(t)\right), \tag{2.34}$$

where $\lambda : [t_{k-1},\ t_k] \to \mathbb{R}^n$ is the costate. Following the analysis of the calculus of variations (Bryson, 2018), a minimizer $d_f^\star$ of the cost (2.28) has to satisfy the stationarity condition on the Hamiltonian (2.34),

$$\frac{\partial H(\bar{x}_f(t), d_f(t), \lambda(t))}{\partial d_f(t)} = 0 \Longrightarrow d_f^\star(t) = -\Delta \cdot \phi(D^{\mathrm{T}}\lambda(t)), \tag{2.35}$$

where $\phi(\cdot)$ is applied elementwise. In addition, the corresponding costate needs to evolve dynamically as

$$-\dot{\lambda}(t) = \frac{\partial H(\bar{x}_f(t), d_f(t), \lambda(t))}{\partial \bar{x}_f(t)} \Longrightarrow \dot{\lambda}(t) = -A^{\mathrm{T}}\lambda(t). \tag{2.36}$$

Integrating (2.36) backward over $t \in [t_{k-1},\ t_k]$, we obtain $\lambda(t) = e^{A^{\mathrm{T}}(t_k-t)}\lambda(t_k)$, $t \in [t_{k-1},\ t_k]$. Plugging this costate into (2.35) and setting $d_f(t) = d_f^\star(t)$, dynamics (2.29) for $t \in [t_{k-1},\ t_k]$ are given by

$$\dot{\bar{x}}_f(t) = A\bar{x}_f(t) - \Delta D\phi(D^{\mathrm{T}} e^{A^{\mathrm{T}}(t_k-t)}\lambda(t_k)). \tag{2.37}$$

Integrating (2.37) over $t \in [t_{k-1},\ t_k]$ and taking into account the boundary constraints (2.30)–(2.31) yield

$$\bar{x}_{fk} = -\int_{t_{k-1}}^{t_k} \Delta e^{A(t_k-\tau)} D\phi\left(D^{\mathrm{T}} e^{A^{\mathrm{T}}(t_k-\tau)}\lambda(t_k)\right) \mathrm{d}\tau, \tag{2.38}$$

which is equivalent to (2.33). Before concluding, note that conditions (2.35)–(2.36) are only necessary for optimality. However, dynamics (2.29) are linear with respect to $\bar{x}_f(t)$ and $d_f(t)$, and the equality constraints (2.30)–(2.31) are also linear. In addition, the Hessian matrix

$$\frac{\partial^2 r(\bar{x}_f(t), d_f(t))}{\partial d_f(t)^2} = \mathrm{diag}\left(\frac{\pi}{2\Delta}\sec^2\left(\frac{\pi}{2\Delta}d_{f1}(t)\right), \ldots, \frac{\pi}{2\Delta}\sec^2\left(\frac{\pi}{2\Delta}d_{fm}(t)\right)\right)$$

is positive definite over the feasible set of disturbances for this optimization problem, which are bounded in terms of their infinity norm by Δ. Thus $r(\bar{x}_f(t), d_f(t))$ is jointly convex over $\bar{x}_f(t)$ and $d_f(t)$. Hence the Mangasarian sufficient conditions (Mangasarian, 1966) are satisfied, making (2.35)–(2.36) both necessary and sufficient conditions for optimality. Therefore, since a feasible solution to problem (2.28)–(2.31) exists if and only if an admissible disturbance exists for system (2.24) over $t \in [t_{k-1},\ t_k]$, a solution to Eq. (2.38) with respect to the constant vector $\lambda(t_k) \in \mathbb{R}^n$ exists if and only if system (2.24) is not under a detectable attack over $t \in [t_{k-1},\ t_k]$. □

Remark 2.6. The condition of Lemma 2.2 has the advantage that it is both sufficient and necessary for the existence of a detectable attack. Therefore a detection mechanism relying on the verification of (2.33) will not suffer from false positives or negatives. However, Eq. (2.33) is nonlinear with respect to the constant vector v, and verifying whether such a vector that satisfies this equation exists is not straightforward. ■

2.3.3 An ϵ-close detection mechanism

Verifying the existence of solutions to (2.33) is sufficient and necessary to detect actuator attacks, but doing can be relatively difficult owing to the nonlinear nature of (2.33). To deal this issue, we can derive an “approximate” detection mechanism by solving a relaxed version of the constrained optimization problem formed in Lemma 2.2.

Theorem 2.4. *Consider, for $t \in [t_{k-1},\ t_k]$, $k \in \mathbb{N}_+$, the parametric system*

$$\dot{\bar{x}}_f(t;\ \epsilon) = A\bar{x}_f(t;\ \epsilon) - \Delta D\phi\left(\frac{1}{\epsilon}D^{\mathrm{T}}\mu(t;\ \epsilon)\right), \tag{2.39}$$

$$\dot{\mu}(t;\ \epsilon) = -A^{\mathrm{T}}\mu(t;\ \epsilon), \tag{2.40}$$

along with the boundary conditions

$$\bar{x}_f(t_{k-1};\ \epsilon) = 0,\ \ \mu(t_k;\ \epsilon) = S\left(\bar{x}_f(t_k;\ \epsilon) - \bar{x}_{fk}\right), \tag{2.41}$$

where $\epsilon > 0$, and $S \succ 0$ is a symmetric matrix. Let $\sigma(\bar{x}_f(t_k;\ \epsilon)) = \frac{1}{2}(\bar{x}_f(t_k;\ \epsilon) - \bar{x}_{fk})^{\mathrm{T}}S(\bar{x}_f(t_k;\ \epsilon) - \bar{x}_{fk})$. Then system (2.24) is under a detectable attack over $t \in [t_{k-1},\ t_k]$ if and only if

$$\lim_{\epsilon\to 0^+} \sigma\left(\bar{x}_f(t_k;\ \epsilon)\right) > 0. \tag{2.42}$$

Proof. Define the ϵ-relaxed version of the optimization problem (2.28)–(2.31) as

$$\begin{aligned}
\min_{d_f(\cdot;\ \epsilon):[t_{k-1},\ t_k]\to\mathbb{R}^{m_d}} \quad & J\left(d_f(\cdot;\ \epsilon);\ \epsilon\right) = \int_{t_{k-1}}^{t_k} \epsilon\cdot r(\bar{x}_f(\tau;\ \epsilon), d_f(\tau;\ \epsilon))\mathrm{d}\tau \\
& + \sigma(\bar{x}_f(t_k;\ \epsilon)), && (2.43)\\
\text{s.t.} \quad & \dot{\bar{x}}_f(t;\ \epsilon) = A\bar{x}_f(t;\ \epsilon) + Dd_f(t;\ \epsilon),\ \ t\in[t_{k-1},\ t_k], && (2.44)\\
& \bar{x}_f(t_{k-1};\ \epsilon) = 0. && (2.45)
\end{aligned}$$

Note that we have replaced the terminal state constraint (2.31) with the terminal cost $\sigma(\bar{x}_f(t_k;\ \epsilon))$, and thus a feasible solution to this optimization always exists. Define the new Hamiltonian as

$$\begin{aligned}
H_n(\bar{x}_f(t;\ \epsilon), d_f(t;\ \epsilon), \mu(t;\ \epsilon);\ \epsilon) := & \sum_{i=1}^{m_d} \epsilon \int_0^{d_{fi}(t;\ \epsilon)} \phi^{-1}\left(\frac{w}{\Delta}\right)\mathrm{d}w \\
& + \mu^{\mathrm{T}}(t;\ \epsilon)\left(A\bar{x}_f(t;\ \epsilon) + Dd_f(t;\ \epsilon)\right),
\end{aligned}$$

where $\mu(\cdot;\ \epsilon) : [t_{k-1},\ t_k] \to \mathbb{R}^n$ is the costate. Employing the same necessary conditions for optimality, a minimizer $d_f^\star$ of (2.43) satisfies

$$\frac{\partial H_n(\bar{x}_f(t;\ \epsilon), d_f(t;\ \epsilon), \mu(t;\ \epsilon);\ \epsilon)}{\partial d_f(t;\ \epsilon)} = 0 \Longrightarrow d_f^\star(t;\ \epsilon) = -\Delta\phi\left(\frac{1}{\epsilon}D^{\mathrm{T}}\mu(t;\ \epsilon)\right), \tag{2.46}$$

$$
-\dot{\mu}(t;\ \epsilon)=\frac{\partial H_n(\bar{x}_f(t;\ \epsilon),d_f(t;\ \epsilon),\mu(t;\ \epsilon);\ \epsilon)}{\partial \bar{x}_f(t;\ \epsilon)}\Longrightarrow \dot{\mu}(t;\ \epsilon)=-A^{\mathrm{T}}\mu(t;\ \epsilon),
\tag{2.47}
$$

whereas the transversality condition yields

$$
\mu(t_k;\ \epsilon)=\frac{\partial \sigma(\bar{x}_f(t_k;\ \epsilon))}{\partial \bar{x}_f(t_k;\ \epsilon)}=S(\bar{x}_f(t_k;\ \epsilon)-\bar{x}_{fk}).
\tag{2.48}
$$

Setting $d_f(t;\ \epsilon)=d_f^{\star}(t;\ \epsilon)$, Eqs. (2.44)–(2.45) and (2.46)–(2.48) become equivalent to (2.39)–(2.41). In addition, since the terminal cost $\sigma(\bar{x}_f(t_k;\ \epsilon))$ is convex with respect to $\bar{x}_f(t_k;\ \epsilon)$, following the same arguments as in Lemma 2.2, the necessary conditions for optimality are sufficient, and the solution given by (2.46)–(2.48) subject to (2.44)–(2.45) is indeed a minimizer of the cost (2.43).

"If" part: Suppose now than an admissible disturbance exists over the interval $t\in[t_{k-1},\ t_k]$ for system (2.24), and denote it as d_a. Then, if we allow $d_f(t;\ \epsilon)=d_a(t)$ in (2.44) for all $t\in[t_{k-1},\ t_k]$, then due to Fact 2.1, we have $\bar{x}_f(t_k;\ \epsilon)=\bar{x}_{fk}$, and thus $\sigma(\bar{x}_f(t_k;\ \epsilon))=0$. In addition, since for an admissible disturbance, we have $\sup_{t\in[t_{k-1},t_k]}\left\|d_a(t)\right\|<\Delta$, if we let $d_f(t;\ \epsilon)=d_a(t)$ for $t\in[t_{k-1},\ t_k]$ in (2.44), then we get $r(\bar{x}_f(t;\ \epsilon),d_a(t))<\infty$ for all $t\in[t_{k-1},\ t_k]$. Overall, we derive

$$
\lim_{\epsilon\to0^+}J(d_a(\cdot);\ \epsilon)=\lim_{\epsilon\to0^+}\int_{t_{k-1}}^{t_k}\epsilon\cdot r(\bar{x}_f(\tau;\ \epsilon),d_a(\tau))\mathrm{d}\tau+\sigma(\bar{x}_f(t_k;\ \epsilon))=0.
$$

However, since (2.46)–(2.48) is a minimizer of the optimization problem (2.43)–(2.45), we have that $J(d_f^{\star}(\cdot;\ \epsilon);\ \epsilon)\leq J(d_a(\cdot);\ \epsilon)$. Since $J(\cdot;\ \epsilon)\geq0$, if we let $d_f=d_f^{\star}$ in (2.44), then we conclude that

$$
\lim_{\epsilon\to0^+}J(d_f^{\star}(\cdot;\ \epsilon);\ \epsilon)=0\Longrightarrow\lim_{\epsilon\to0^+}\sigma(\bar{x}_f(t_k;\ \epsilon))=0.
$$

"Only If" part: Finally, suppose that an admissible disturbance for system (2.24) over $t\in[t_{k-1},\ t_k]$ does not exist. Then $\bar{x}_f(t_k;\ \epsilon)\neq\bar{x}_{fk}$ for all $\epsilon>0$ and all feasible solutions d_f of (2.43)–(2.45); otherwise, an admissible disturbance will exist by its definition. Thus, since $S\succ0$, over the optimal trajectories, we have $\sigma(\bar{x}_f(t_k;\ \epsilon))\neq0$ and, consequently, also $\lim_{\epsilon\to0^+}\sigma(\bar{x}_f(t_k;\ \epsilon))\neq0$, which concludes the theorem since $\sigma(\cdot)$ cannot be negative. □

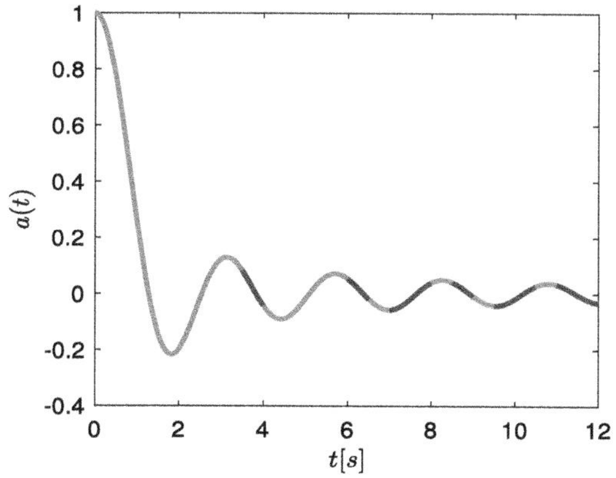

Figure 2.11 The evolution of the attacked input. The detectable parts are in red (mid gray in print version), as declared by the ϵ-close detector.

2.3.4 Simulation results

To showcase the applicability of the method, we conduct two simulations on the Aero-Data Model in Research Environment (ADMIRE) benchmark aircraft of Yu and Jiang (2012). The aircraft system matrices are

$$A = \begin{bmatrix} -1.0649 & 0.0034 & -0.0000 & 0.9728 & 0.0000 \\ 0.0000 & -0.2492 & 0.0656 & -0.0000 & -0.9879 \\ 0.0000 & -22.5462 & -2.0457 & -0.0000 & 0.5432 \\ 8.1633 & -0.0057 & -0.0000 & -1.0478 & 0.0000 \\ 0.0000 & 1.7970 & -0.1096 & 0.0000 & -0.4357 \end{bmatrix},$$

$$D = \begin{bmatrix} -0.0072 \\ 0.0000 \\ 0.0000 \\ 0.0551 \\ -0.0000 \end{bmatrix},$$

$$B = \begin{bmatrix} -0.0062 & -0.0062 & -0.0709 & -0.1172 & -0.1172 & -0.0709 & 0.0003 \\ -0.0072 & 0.0072 & 0.0039 & 0.0188 & -0.0188 & -0.0039 & 0.0627 \\ 1.2456 & -1.2456 & -10.6058 & -9.2345 & 9.2345 & 10.6058 & 5.3223 \\ 2.7172 & 2.7172 & -2.4724 & -4.0101 & -4.0101 & -2.4724 & 0.0108 \\ -0.7497 & 0.7497 & -0.4923 & -1.1415 & 1.1415 & 0.4923 & -3.7367 \end{bmatrix}.$$

We assume that the known bound of the disturbance is $\Delta = 0.06$ and that the state $x(t)$ is available for measurement only at the time instants $t_k = 0.5k$ [s], $k \in \mathbb{N}$. We choose the real disturbance as $d(t) = 0.05 \cdot \sin(\pi t)$

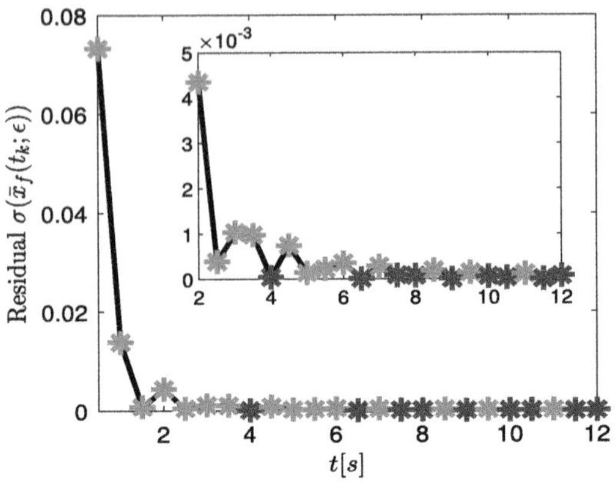

Figure 2.12 The evolution of the residual $\sigma(\bar{x}_f(t_k;\ \epsilon))$ of the ϵ-close detector for the aircraft under an actuator attack. Given that $\sigma(\bar{x}_f(t_k;\ \epsilon)) \geq \delta = 10^{-4}$, an attack is declared [in red (mid gray in print version)].

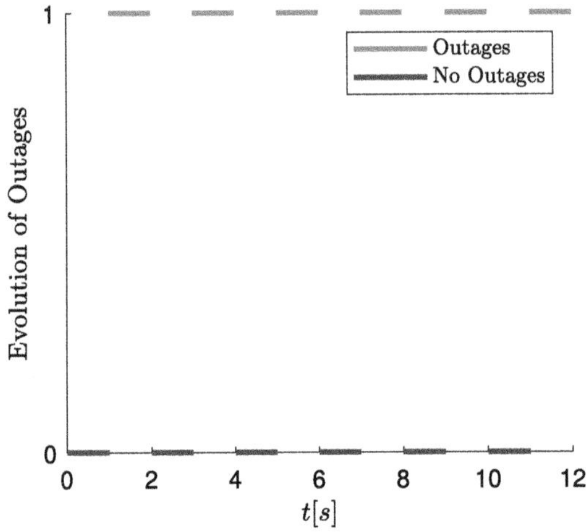

Figure 2.13 The evolution of the actuator outages.

and the control input $u(t)$ as the solution of a min-max model predictive control problem. Finally, we pick $\epsilon = 10^{-5}$, $\delta = 10^{-4}$, $S = I_5$, and an initial condition equal to $\begin{bmatrix} 0.1\ 0.1\ 0.1\ 0.1\ 0.1 \end{bmatrix}^{\mathrm{T}}$. The remaining parameters are picked as in the previous example.

For the first simulation, we consider that the first actuator of the system suffers from an attack, so that $K = \begin{bmatrix} -0.0062\ -0.0072\ 1.2456\ 2.7172\ -0.7497 \end{bmatrix}^{\mathrm{T}}$, and we choose the attack signal as $a(t) = \mathrm{sinc}(\frac{\pi t}{4})$. The attack and the dec-

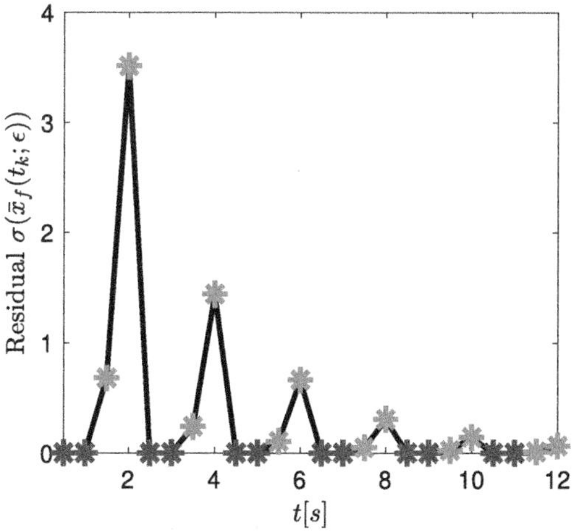

Figure 2.14 The evolution of the residual $\sigma(\bar{x}_f(t_k;\ \epsilon))$ of the ϵ-close detector for the aircraft under actuator outages. Given that $\sigma(\bar{x}_f(t_k;\ \epsilon)) \geq \delta = 10^{-4}$, an attack is declared [in red (mid gray in print version)].

larations of the ϵ-close detection mechanism are depicted in Figs. 2.11 and 2.12. As we can notice, the attack is detectable as long as it takes relatively high values. As the attack vanishes, it eventually blends with the disturbance, becoming indistinguishable. The mean running time for the boundary value problem solved at each iteration was 81 ms. Although this time is satisfactory for the problem at hand, it can be significantly reduced by using a compiled programming language or a better processing unit.

For the second simulation, we consider that all the actuators of the system suffer from periodic communication failures, hence being removed from the system in a periodic fashion. It then follows that $K = B$, $a(t) = -u(t)$ every time that the actuator outages take place, and $a(t) = 0$ whenever there is no outage. The results are shown in Figs. 2.13 and 2.14, where it is evident that all the outages have been detected. The mean running time for the boundary-value problem solved at each iteration was 83 ms.

CHAPTER THREE

Redundancy-based defense

3.1. Moving target defense

In designing a defense mechanism employing the principles of unpredictability for CPS, we revisit a linear time-invariant continuous-time model of its behavior:

$$\begin{aligned} \dot{x}(t) &= Ax(t) + Bu_a(t), \ \ t \geq 0, \\ y(t) &= C_a(t)x(t), \end{aligned} \tag{3.1}$$

where we remind the reader that $x(t) \in \mathbb{R}^n$ is the state, $u_a(t) \in \mathbb{R}^m$ is the potentially attacked input of the system, $y(t) \in \mathbb{R}^p$ is the output, $A \in \mathbb{R}^{n\times n}$ is the plant matrix, $B \in \mathbb{R}^{n\times m}$ is the input matrix, and $C_a(t) \in \mathbb{R}^{p\times n}$ is the potentially attacked output matrix.

To better highlight the redundancy properties of the CPS, we rewrite (3.1) as

$$\begin{aligned} \dot{x}(t) &= Ax(t) + \sum_{i=1}^{m} b^i u^i(t), \\ y_j(t) &= c_j(t)x(t) \ , j \in \{1, \ldots, p\}, \end{aligned}$$

where b^i is a column vector corresponding to the ith actuator, $u^i \in \mathbb{R}$ is the value of the input signal associated with this actuator, and y_j is the output given by a specific sensor c_j corresponding to the jth row of the output matrix.

The potentially compromised control input of (3.1) will be of the following form:

$$u_a(t) = \rho(t)u(t), \ \ t \geq 0, \tag{3.2}$$

where $\rho(t) = \text{diag}(\rho_{ii}(t), i \in \{1, \ldots, m\})$ is a time-varying actuator attack parameter controlled by an adversary, and $u(t) \in \mathbb{R}^m$ is the non-attacked control input. The output matrix of the system can be undermined by a signal $\rho^s(t)$ as

$$C_a(t) = \rho^s(t)C, \ \ t \geq 0, \tag{3.3}$$

where $\rho^s(t)$ is a diagonal matrix controlled by the attacker, and $C \in \mathbb{R}^{p\times n}$ is the non-attacked output matrix.

Control and Game Theoretic Methods for Cyber-Physical Security
https://doi.org/10.1016/B978-0-44-315408-9.00009-9

Remark 3.1. Note that the focus of our work is on the components of the CPS that can be modeled utilizing control-theoretic techniques. Although there are attack angles that can affect the software that implements the proposed intrusion detection algorithms, those lie beyond the scope of our research. It is assumed that the computing elements are equipped with appropriate security measures, such as encryption mechanisms (McLaughlin et al., 2010). On the other hand, our approach considers attacks that leverage the dual nature of CPS; hence we develop methods that take into account the cyber-components that interact with the physics of the systems, i.e., sensors and actuators. ■

Assumption 3.1. To offer a greater degree of freedom for deception purposes and to mitigate the effect of potential attacks, we will consider systems with redundant actuating and sensing components. ■

Assumption 3.2. The system actuators are not compromised over a time interval $\tau \in [t_1, t_2]$ if and only if $\rho_{ii}(\tau) = 1$, $i \in \{1, \ldots, m\}$. Similarly, we consider the sensors as secure if and only if for all τ, $\rho^s_{jj}(\tau) = 1$, $j \in \{1, \ldots, p\}$. Signals (3.2) and (3.3) are assumed to be integrable over any closed time interval $[t_1, t_2]$, $0 \leq t_1 < t_2$. ■

Remark 3.2. The assumption on the local integrability of the adversarial signals allows us to consider a variety of realistic attack scenarios, such as impulses and other discontinuous signals, or constant bias injection, which is locally, but not globally, integrable. The underlying restriction on the signal excludes attacks that have infinite value on a specific time interval, which is a practical assumption on the adversarial capabilities. ■

Assumption 3.3. The attacker is not able to compromise all the actuators and sensors at once. Therefore $\text{supp}(\rho) < m$ and $\text{supp}(\rho^s) < p$. ■

Remark 3.3. Note that our formulation will make no assumptions on the structure, boundedness, and other Lipschitz continuity properties of the attacker's signal. Furthermore, attacks of the form (3.2) and (3.3), due to their time-varying nature, can describe a wide range of attacks, including additive and multiplicative attacks. ■

We are thus interested in designing both reactive and proactive defense mechanisms that will operate well in the absence of attackers and will detect and mitigate attacks. Having presented our methods on detection, this chapter focuses on the design of proactive methods of security. Thus, the developed algorithms disregard the various attack signals to the system and operate under the ideal dynamics. We will initially focus our attention on the case of actuator attacks. Throughout this section, we assume full state feedback. Let $\mathcal{B}$ denote the set containing the actuators of (3.1) by the vectors b^i, $i \in \{1, \ldots, m\}$. The power set of $\mathcal{B}$, denoted as $2^{\mathcal{B}}$, contains all possible combinations of the actuators acting on (3.1). Each of these combinations is expressed by the input matrix B_j, $j \in \{1, \ldots, 2^m\}$, whose columns are the appropriate vectors b^i.

The set of the candidate actuating modes $\mathcal{B}_c$ is defined as the set of the actuator combinations that renders system (3.1) fully controllable, i.e.,

$$\mathcal{B}_c = \left\{ B_j \in 2^{\mathcal{B}} : \mathrm{rank}\left(\begin{bmatrix} B_j & AB_j & \ldots & A^{n-1}B_j \end{bmatrix} \right) = n \right\}. \tag{3.4}$$

System (3.1) assuming full state-feedback with the actuating mode B_i can be rewritten as

$$\dot{x} = Ax + B_i u_i, \ i \in \{1, \ldots, m\}, \ t \geq 0, \tag{3.5}$$

where u_i is the control signal vector of appropriate dimension corresponding to the specific actuating mode.

Remark 3.4. Note that we do not require different actuating modes to share common actuators. Moreover, while a single actuating mechanism might be able to control a system, two different – less potent – mechanisms might need to work cooperatively to control the same system. All these modes will belong to the set described in (3.4). ■

3.1.1 Optimal controller design

For each actuating operating mode B_i, $i \in \{1, \ldots, m\}$, we denote the candidate control law as $u_i(t)$.

We are interested in deriving optimal controllers for each of these modes by utilizing well-known optimal control approaches (Lewis et al., 2012). Toward that, we are interested in solving the following optimization:

$$V_i^\star(x(t_0)) = \min_{u_i} \int_{t_0}^{\infty} r_i(x, u_i)\mathrm{d}\tau \equiv \min_{u_i} \int_{t_0}^{\infty} (x^{\mathrm{T}} Q_i x + u_i^{\mathrm{T}} R_i u_i)\mathrm{d}\tau \ \ \forall x(t_0), \tag{3.6}$$

given (3.5), where $Q_i \succ 0$, $R_i \succ 0$, $i \in \{1, \ldots, \mathrm{card}(\mathcal{B}_c)\}$.

Assumption 3.4. Each pair $(A, \sqrt{Q_i})$ is observable. ■

The Hamiltonian associated with (3.5) and (3.6) is

$$H_i(x, u_i, \nabla V_i) = \nabla V_i^{\mathrm{T}}(Ax + B_i u_i) + x^{\mathrm{T}} Q_i x + u_i^{\mathrm{T}} R_i u_i \quad \forall x, u_i,$$

with value function V_i, not necessarily optimal.

Applying the stationarity conditions $\frac{\partial H_i(x, u_i, \nabla V_i)}{\partial u_i} = 0$ yields

$$u_i = -R_i^{-1} B_i^{\mathrm{T}} \nabla V_i. \tag{3.7}$$

The optimal value functions $V_i^\star$ must satisfy the following HJB equation:

$$x^{\mathrm{T}} Q_i x + \nabla V_i^{\star\mathrm{T}} Ax - \frac{1}{2} \nabla V_i^{\star\mathrm{T}} B_i R_i^{-1} B_i^{\mathrm{T}} \nabla V_i^\star = 0. \tag{3.8}$$

Since all the systems described by (3.5) are linear and the cost given by (3.6) is quadratic, all the value functions will be quadratic in the state x, i.e., $V_i^\star(x) = x^{\mathrm{T}} P_i x,\ P_i \succ 0$. Substituting this expression into (3.8) and the resulting optimal value function into (3.7) yields the feedback controller with optimal gain K_i:

$$u_i^\star(x) = -K_i x := -R_i^{-1} B_i^{\mathrm{T}} P_i x \quad \forall x,$$

where P_i are the solutions to the following Riccati equations:

$$A^{\mathrm{T}} P_i + P_i A - P_i B_i R_i^{-1} B_i^{\mathrm{T}} P_i + Q_i = 0. \tag{3.9}$$

We introduce the set $\mathcal{K}$ containing all $K_i,\ i \in \{1, \ldots, m\}$, with the understanding that $\mathrm{card}(\mathcal{K}) = \mathrm{card}(\mathcal{B}_c)$.

For ease of exposition, with some abuse of notation, we will use K_i to refer to both the optimal controller with this gain and its corresponding index.

Fact 3.1. Due to (3.4) and Assumption 3.4, for each B_i, the solution exists and is unique. ■

Fact 3.2. Each K_i with input given by (3.7) guarantees that (3.1) has an asymptotically stable equilibrium point. ■

3.1.2 Switching-based MTD framework

We will now develop a framework to facilitate the deception of potential attackers based on the principles of MTD.

Maximization of unpredictability

To formally define the switching law, we need to introduce a *probability* $\mathbf{p}$, which denotes the probability that each controller K_i is active.

To incorporate ideas from the framework of MTD, we propose a switching rule that optimizes over the minimum cost that each controller is able to attain, as well as an unpredictability term quantified by the information entropy produced by the switching probability $\mathbf{p}$. This way, we will achieve the desired trade-off between overall optimality and unpredictability. The use of the information entropy is a standard practice in MTD design (Okhravi et al., 2014).

Suppose that (3.1) is controlled by $N = \text{card}(\mathcal{K})$ candidate controllers with an associated cost given by (3.6). The probability p_i that each controller K_i is active is given by

$$p_i = e^{\left(-\frac{V_i^\star}{\epsilon}-1-\epsilon\log\left(e^{-1}\sum_{i=1}^{N} e^{\frac{V_i^\star}{\epsilon}}\right)\right)}, \tag{3.10}$$

where $\epsilon \in \mathbb{R}^+$ is the weight of unpredictability during the optimization process.

To derive (3.10), we formulate the following optimization problem:

$$\min_{\mathbf{p}} \left(\mathbf{V}^{\star\mathrm{T}}\mathbf{p} - \epsilon\mathcal{H}(\mathbf{p})\right)$$
$$\text{subject to } \|\mathbf{p}\|_1 = 1 \text{ and } \mathbf{p} \succeq 0,$$

where $\mathbf{p} \in \mathbb{R}^N$ is the vector of probabilities constrained in the simplex, $\mathbf{V}^\star := [V_1^\star \ \dots \ V_N^\star]^\mathrm{T} = [x(t_0)^\mathrm{T} P_1 x(t_0) \ \dots \ x(t_0)^\mathrm{T} P_N x(t_0)]^\mathrm{T}$ is the column vector containing the value function of each candidate controller, and $\mathcal{H}(\mathbf{p}) = -\mathbf{p}^\mathrm{T}\log(\mathbf{p})$ is the information entropy produced by the probability.

Remark 3.5. The choice of this objective function allows us to combine the two required specifications. The linear term $\mathbf{V}^{\star\mathrm{T}}\mathbf{p}$ penalizes the deviations from the overall optimal controller, whereas the entropy term $\mathcal{H}(\mathbf{p})$ penalizes the use of a single controller throughout the operation of the system. The result is a compromise specified by the optimization weight ϵ. ■

Furthermore, for the decision vector $\mathbf{p}$ to constitute a point in the probability simplex, we constrain it to the non-negative orthant (i.e., $p_i \leq 0 \ \forall i \in \{1, \dots, N\}$), and we require its l_1 norm to satisfy $\|p\|_1 = \sum_{i=1}^{N} \|p_i\| = 1$.

The entropy of a probability is a concave function (Cover and Thomas, 2012), and therefore the cost index, being a sum of a linear function of the probability and the negative entropy, is convex. Thus we can define the Lagrangian of the optimization problem as

$$\begin{aligned} L &= \mathbf{V}^{\star \mathrm{T}}\mathbf{p} - \epsilon \mathcal{H}(\mathbf{p}) + \lambda(\mathbf{1}^{\mathrm{T}}\mathbf{p} - 1) + \boldsymbol{\beta}^{\mathrm{T}}\mathbf{p} \\ &= \mathbf{V}^{\star \mathrm{T}}\mathbf{p} + \epsilon \mathbf{p}^{\mathrm{T}} \log(\mathbf{p}) + \lambda(\mathbf{1}^{\mathrm{T}}\mathbf{p} - 1) + \boldsymbol{\beta}^{\mathrm{T}}\mathbf{p}, \end{aligned}$$

where $\mathbf{1}$ denotes the vector consisting of ones, and λ and $\boldsymbol{\beta}$ are the Karush–Kuhn–Tucker (KKT) multipliers.

The KKT conditions for the problem are

$$\nabla_{\mathbf{p}} L = \mathbf{V}^{\star} + \epsilon \mathbf{1} + \epsilon \log(\mathbf{p}) + \lambda \mathbf{1} + \boldsymbol{\beta},$$

and the complementarity conditions for the optimal solution $\mathbf{p}^{\star}$ are

$$\boldsymbol{\beta}^{\mathrm{T}}\mathbf{p}^{\star} = 0.$$

If there exists i for which $p_i = 0$, then the term $\log(p_i)$ will be undefined. Consequently, for the optimization problem to be feasible, one of the following two conditions need to hold:

- $\epsilon \log(p_i) = 0 \ \forall i \Rightarrow \epsilon = 0 \Rightarrow \mathbf{p}^{\star} = \left[\mathbf{0}_{i-1} \ldots 1 \ldots \mathbf{0}_{N-i}\right]^{\mathrm{T}}$ where the K_i controller is the one with an overall less cost, and
- $\boldsymbol{\beta} = 0$.

Consider now the nontrivial case $\boldsymbol{\beta} = 0$, which yields

$$\nabla_{\mathbf{p}} L = \mathbf{V}^{\star} + \epsilon \log(\mathbf{p}) + \epsilon \mathbf{1} + \lambda \mathbf{1} = 0.$$

The N equations for each controller are independent, leading to the following system of equations:

$$V_i^{\star} + \epsilon \log(p_i) + \epsilon + \lambda = 0, \ \ i \in \{1, \ldots, N\}.$$

Solving now for the optimal probabilities p_i yields

$$p_i = e^{\left(-\frac{V_i^{\star}}{\epsilon} - \frac{\lambda}{\epsilon} - 1\right)}, \ \ i \in \{1, \ldots, N\}. \tag{3.11}$$

Taking into account that

$$\|\mathbf{p}\|_1 = 1 \Rightarrow \sum_{i=1}^{N} p_i = 1 \Rightarrow \sum_{i=1}^{N} e^{\left(-\frac{V_i^{\star}}{\epsilon} - \frac{\lambda}{\epsilon} - 1\right)} = 1$$

and solving for λ yield

$$\lambda = \epsilon \log\left(e^{-1}\sum_{i=1}^{N} e^{\left(-\frac{V_i^{\star}}{\epsilon}\right)}\right). \tag{3.12}$$

Substituting (3.12) into (3.11) provides the required result.

Switching-based MTD scheme

To analyze the behavior of the system under the proposed MTD framework, we will formulate a switched system consisting of the different operating modes.

First, we introduce the switching signal $\sigma(t) = i,\ i \in \{1, \ldots, \mathrm{card}(\mathcal{K})\}$, which denotes the active controller as a function of time. This way, the system is

$$\dot{x}(t) = \tilde{A}_{\sigma(t)} x(t), \tag{3.13}$$

where $\tilde{A}_{\sigma(t)} := A - B_{\sigma(t)} R_{\sigma(t)}^{-1} B_{\sigma(t)}^{\mathrm{T}} P_{\sigma(t)}$ denotes the closed-loop subsystem with the controller $K_{\sigma(t)}$ active.

Remark 3.6. Since the actual switching sequence is different under the designer's choice for unpredictability, we will constrain the switching signal to have a predefined average dwell time. This way, the stability of the overall system will be independent of the result of optimization. Intuitively, as was initial shown in Hespanha and Morse (1999), a system with stable subsystems is stable if the switching is slow enough on average. ■

Definition 3.1. A switching signal has an average dwell time τ_D if over any time interval $[t, T],\ T \geq t$, the number of switches $S(T, t)$ is bounded above as

$$S(T, t) \leq S_0 + \frac{T - t}{\tau_D},$$

where S_0 is an arbitrary chatter bound, and τ_D is the dwell time. ■

Theorem 3.1. *Consider system* (3.1) *in the absence of attacks. The switched system defined by the piecewise continuous switching signal* $\sigma(t) = i,\ i \in \{1, \ldots, \mathit{card}(\mathcal{K})\}$, *with active controller* K_i *given by* (3.7) *and continuous flow given by* (3.5) *has an asymptotically stable equilibrium point for every switching* *signal $\sigma(t)$ if the average dwell time is bounded by*

$$\tau_D > \frac{\log\left(\max_{q,p\in\{1,\ldots,card(\mathcal{K})\}} \frac{\bar{\lambda}(P_p)}{\underline{\lambda}(P_q)}\right)}{\min_{p\in\{1,\ldots,card(\mathcal{K})\}} \frac{\underline{\lambda}(Q_p+P_pB_pR_p^{-1}B_p^{T}P_p)}{\underline{\lambda}(P_p)}} \tag{3.14}$$

with an arbitrary chatter bound $S_0 > 0$.

Proof. For each $i \in \{1, \ldots, \mathrm{card}(\mathcal{K})\}$, following Liberzon (2012), for each subsystem, we choose the Lyapunov function

$$\mathcal{V}_i(x) = x^{T} P_i x \;\; \forall x,$$

where P_i is the solution to the Riccati equation (3.9). The Lyapunov functions are positive definite and radially unbounded for all x. According to the Rayleigh–Ritz inequality for symmetric matrices, we have

$$\underline{\lambda}(P_i)\|x\|^2 \leq x^{T} P_i x = \mathcal{V}_i(x) \leq \bar{\lambda}(P_i)\|x\|^2. \tag{3.15}$$

The time derivative of $\mathcal{V}_i(x)$ along the solutions of the trajectory of the corresponding subsystem is

$$\begin{aligned}\dot{\mathcal{V}}_i(x) &= \dot{x}^{T} P_i x + x^{T} P_i \dot{x}\\ &= x^{T}(A - B_iR_i^{-1}B_i^{T}P_i)^{T}P_i x + xP(A - B_iR_i^{-1}B_i^{T}P_i)x\\ &= x^{T}(A^{T}P_i - P_iB_iR_i^{-1}B_i^{T}P_i + P_iA - P_iB_iR_i^{-1}B_i^{T}P_i)x.\end{aligned}$$

Taking into account (3.9) and denoting $\bar{H}_i := Q_i + P_iB_iR_i^{-1}B_i^{T}P_i \succ 0$, $i \in \{1, \ldots, \mathrm{card}(\mathcal{K})\}$, yields

$$\dot{\mathcal{V}}_i(x) = -x^{T}\bar{H}_i x.$$

Consequently,

$$\dot{\mathcal{V}}_i(x) \leq -\underline{\lambda}(\bar{H}_i)\|x\|^2. \tag{3.16}$$

Combining now (3.16) with (3.15) and noting that $\underline{\lambda}(P_i)\|x\|^2 \leq \mathcal{V}_i(x) \Rightarrow \|x\|^2 \leq \frac{1}{\underline{\lambda}(P_i)}\mathcal{V}_i(x)$ yield

$$\dot{\mathcal{V}}_i(x) \leq -\frac{\underline{\lambda}(\bar{H}_i)}{\underline{\lambda}(P_i)}\mathcal{V}_i(x). \tag{3.17}$$

For the inequality to hold for arbitrary modes, we have

$$\dot{\mathcal{V}}_i(x) \leq -\min_{i\in\{1,\ldots,\mathrm{card}(\mathcal{K})\}} \frac{\underline{\lambda}(\bar{H}_i)}{\underline{\lambda}(P_i)}\mathcal{V}_i(x).$$

Following similar arguments, we show that for all $p, q \in \{1, \dots, \text{card}(\mathcal{K})\}$,

$$\mathcal{V}_p(x) \le \frac{\bar{\lambda}(P_p)}{\underline{\lambda}(P_q)} \mathcal{V}_q(x).$$

For the inequality to hold for arbitrary pairs of modes, we further write

$$\mathcal{V}_p(x) \le \max_{p,q \in \{1,\dots,\text{card}(\mathcal{K})\}} \frac{\bar{\lambda}(P_p)}{\underline{\lambda}(P_q)} \mathcal{V}_q(x).$$

For ease of exposition, we denote

$$\nu := \min_{i \in \{1,\dots,\text{card}(\mathcal{K})\}} \frac{\underline{\lambda}(\bar{H}_i)}{\underline{\lambda}(P_i)} \quad \text{and} \quad \mu := \max_{p,q \in \{1,\dots,\text{card}(\mathcal{K})\}} \frac{\bar{\lambda}(P_p)}{\underline{\lambda}(P_q)}.$$

Without loss of generality, we will consider that the switched system is evolving on the time interval $[0, t_f]$. Denote by $S(t_f, 0)$ the number of switches over this interval, which takes place at times t_i, $i \in [0, S(t_f, 0)]$, with $t_i < t_{i+1}$. The active mode will be the same over any interval $[t_i, t_{i+1}]$, i.e., the switching signal $\sigma(t) = i$ is piecewise constant.

Define the function

$$W(t) = e^{-\nu t} \mathcal{V}_{\sigma(t)}(x(t)). \tag{3.18}$$

Along the solutions of the switched system (3.13) over an interval $t \in [t_i, t_{i+1}]$, the time derivative of (3.18) is

$$\dot{W} = -\nu W + e^{-\nu t} \dot{\mathcal{V}}_{\sigma(t)}(x(t)),$$

which is non-positive due to (3.17). Consequently, the function $W(t)$ is a non-increasing function for all $t \in [t_i, t_{i+1}]$.

At the jump instances t_i, we have

$$\begin{aligned} W(t_{i+1}) &= e^{-\nu t_{i+1}} \mathcal{V}_{\sigma(t_{i+1})}(x(t_{i+1})) \\ &\le \mu e^{-\nu t_{i+1}} \mathcal{V}_{\sigma(t_{i+1})}(x(t_{i+1})) \Rightarrow \\ W(t_{i+1}) &\le \mu e^{-\nu t_i} \mathcal{V}_{\sigma(t_i)}(x(t_i)) = \mu W(t_i), \end{aligned} \tag{3.19}$$

where we used the non-increase of $W(t)$.

Over the whole interval $[0, t_f]$, iterating (3.19) over the $S(0, t_f) - 1$ discontinuities yields

$$W(t_f-) \le \mu^{S[0,t_f]} W(0) \Rightarrow$$

$$
\begin{aligned}
& e^{\nu t_f} \mathcal{V}_{\sigma(t_f-)}(x(t_f)) \leq \mu^{S[0,t_f]} e^{\nu 0} \mathcal{V}_{\sigma(0)}(x(0)) \Rightarrow \\
& e^{\nu t_f} \mathcal{V}_{\sigma(t_f-)}(x(t_f)) \leq \mu^{S[0,t_f]} \mathcal{V}_{\sigma(0)}(x(0)).
\end{aligned} \tag{3.20}
$$

We can rewrite now (3.20) as

$$
\mathcal{V}_{\sigma(t_f-)}(x(t_f)) \leq e^{S_0 \log \mu} e^{\left(\frac{\log \mu}{\tau_D} - \nu\right) t_f} \mathcal{V}_{\sigma(0)}(x(0)).
$$

It is clear that choosing τ_D in a way that satisfies the bound (3.14), the exponential terms are such that $\mathcal{V}_{\sigma(t_f-)}(x(t_f)) \to 0$ as $t_f \to \infty$. Due to (3.15), we can conclude that $x(t_f) \to 0$, which is the required result. □

3.1.3 Proactive and reactive defense for actuator attacks

The described defense method based on unpredictability can be combined with available intrusion detection mechanisms to ensure overall smooth performance of the system in the presence of attacks. As such, we present a way of combining MTD with the Bellman-based intrusion detection method investigated in Chapter 2.

Under safe operation, the system switches between the available modes with MTD to guarantee the stability and maximal unpredictability according to (3.11). If we can find $i : e_i(t_k) \neq 0$, then we can conclude that the ith mode is considered under attack and is isolated. Specifically, the system switches to the controller with the best performance, and the compromised ith mode is taken out of the queue for the MTD switching. The pseudocode for the proactive and reactive defense system is provided in Algorithm 3.1.

Fact 3.3. It has been shown that $p_i > 0$, $i \in \{1, \ldots, \text{card}(\mathcal{K})\}$. Consequently, there exists $t_f^\star$ such that there is $\tau \in [t_0, t_f^\star]$ with $\sigma(\tau) = i$, $i \in \{1, \ldots, \text{card}(\mathcal{K})\}$, and arbitrary $t_0 > 0$. This means that since the probability that all controllers will eventually be active, there is a time interval long enough such that we have already switched through every available controller. ■

Theorem 3.2. *Suppose that* (3.1) *uses the framework of Algorithm* 3.1. *Then the closed-loop system has an asymptotically stable equilibrium point given that the attacker has not compromised all the available controllers, i.e.,* $\mathcal{K} \setminus \mathcal{K}_c \neq \emptyset$, *where* $\mathcal{K}_c$ *is the subset of those controllers that have been compromised by an attacker.*

Proof. We consider a trajectory of the system within the time interval $t \in [t_0, t_f]$, $t_f > t_f^\star$. Denote by $\mathcal{K}_u$ the set of safe controllers and by $\mathcal{K}_c$ the set of compromised ones. Recall that according to Algorithm 3.1 and Theorem 3.2, the controller will stay at a compromised mode during $[t, t+T]$.

Algorithm 3.1 Proactive/Reactive Defense Mechanism for Actuator Attacks.

1: **procedure**
2: Given an initial state $x(t_0)$ and a time window T.
3: Find all permutations of actuators (columns of B) and derive the subset of controllable pairs (A, B_i), denoted by $\mathcal{K}$.
4: **for** $i = 1, \ldots, \text{card}(\mathcal{K})$
5: Compute the optimal feedback gain and Riccati matrices K_i and P_i according to (3.7) and (3.9).
6: Compute the optimal cost of each controller for the given $x(t_0)$.
7: **end for**
8: Solve for the optimal probabilities $p_i^\star$ using (3.10).
9: At $t = t_0$, choose the optimal controller for which $\sigma(t_0) = \arg\min_i \left(x(t_0)^{\mathrm{T}} P_i x(t_0)\right)$.
10: **while** $\sigma(t) = i$ and $t < \tau_D$
11: Compute the integral Bellman error detection signal using (2.5).
12: Propagate the system using (3.5).
13: **end while**
14: Choose the random mode $\sigma(t + \tau_D) = j$ and go to 9.
15: **if** $\|e_i(t_c)\| > 0$
16: Take the ith controller offline.
17: Switch to the controller with the best performance, $\sigma(t_c) = \arg\min_{i \in \mathcal{K} \setminus i} \left(x(t_0)^{\mathrm{T}} P_i x(t_0)\right)$ and go to 9.
18: **end if**
19: **end procedure**

Since the MTD algorithm is constrained by an average dwell time, for any part of the trajectory $t \in [t_k, t_{k+1}]$ where a compromised controller has not been utilized, according to Theorem 3.1, we have

$$\|x(t_{k+1})\| < \|x(t_k)\|. \tag{3.21}$$

We now need to consider those instances where, after detecting a compromised controller K_i, the system immediately switches to another compromised controller. For $\eta \in \mathbb{N}$ subsequent switches to compromised controllers, due to Lemma 2.1, those parts of the trajectory for $t \in [t_k, t_{k+1}] =$

$[t_k, t_k + \eta T]$ are bounded by a positive definite function $\beta_i(\rho, \tau)$ as

$$\begin{aligned}\|x(t_k + \eta T)\| &\leq \beta_{\sigma(t_k+\eta T)}(\rho, T)\|x(t_k + (\eta - 1)T)\| \\ &\leq \beta_{\sigma(t_k+\eta T)}(\rho, T)\beta_{\sigma(t_k+(\eta-1)T)}(\rho, T)\|x(t_k + (\eta - 2)T)\| \\ &\leq \prod_{i=1}^{\eta} \beta_i(\rho, T)\|x(t_k)\|. \end{aligned} \tag{3.22}$$

Furthermore, (3.22) can be upper bounded as

$$\|x(t_k + \eta T)\| \leq \left(\max_{i\in\mathcal{K}_c}(\beta_i(\rho, T))\right)^{\eta} \|x(t_k)\|.$$

Since $\beta_i(\rho(t), T) > 0$ for all i, we can conclude that the parts of the trajectory where the compromised and safe modes are interchanged are upper bounded by the same trajectory driven only by compromised modes, by combining inequalities (3.21) and (3.22).

By Assumption 3.3 the attacker under finite resources is able to compromise a number N of the available controllers, i.e., $\text{card}(\mathcal{K}_c) \leq N$. Using Algorithm 3.1 and Fact 3.3, there is a time $t_p < t_f$ such that all the compromised controllers have been detected by the integral Bellman detector and have been isolated from the switching queue of the MTD.

Consequently, recalling that every closed-loop matrix $\tilde{A}_i = A - B_i K_i$ is Hurwitz, there exist positive numbers $\bar{K}_i$ and a_i such that $\|e^{\tilde{A}_i t}\| \leq \bar{K}_i e^{-a_i t}$, we have

$$\|x(t_f)\| \leq \bar{K}_i e^{-a_i t}\|x(t_p)\| \leq \bar{K}_i e^{-a_i t} \left(\max_{i\in\mathcal{K}_c}(\beta(\rho, T))\right)^{\eta} \|x(t_0)\|.$$

We can see that the remaining trajectory converges to the origin exponentially fast with rate that depends on the slowest safe controller. As a result, if the set of the safe controllers is not empty, then the trajectory is guaranteed to asymptotically go to zero as $t_f \to \infty$. □

3.1.4 Simulation results

Similarly to Chapter 2, we utilize the ADMIRE aircraft as a benchmark system to show the efficacy of our approaches. We remind the reader that ADMIRE has seven redundant actuators and two redundant sensors. Initially, we present results for the problem of controlling the plant in an adversarial environment. Fig. 3.1 shows the switching signal for the MTD framework applied to actuator attacks. We can see that the actuator with

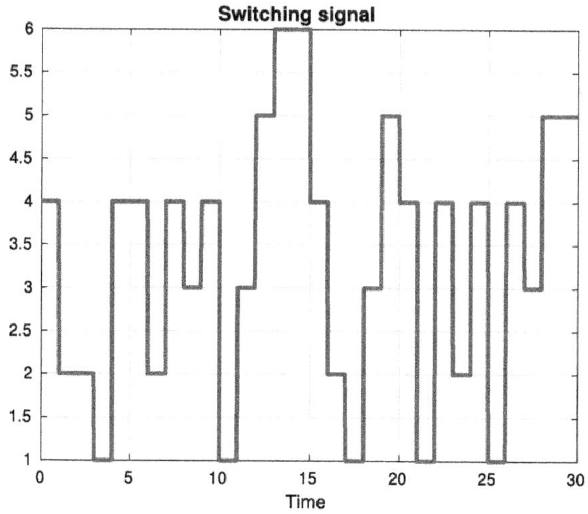

Figure 3.1 The evolution of the MTD switching signal that guarantees actuator proactive security. We can see that the controller with index 4 is preferred since it is the most optimal.

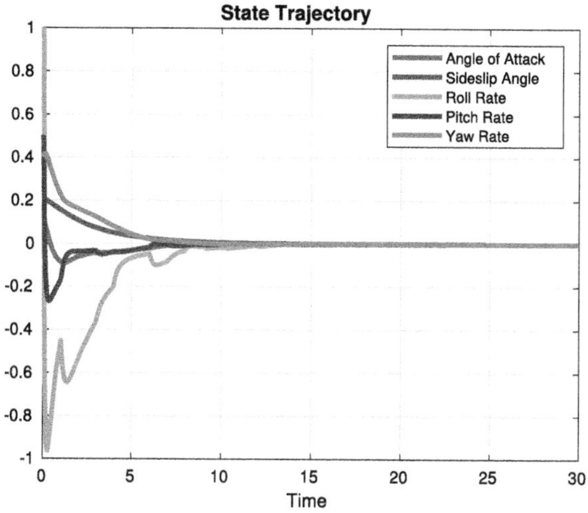

Figure 3.2 The evolution of the MTD state that guarantees actuator proactive security. With an appropriate dwell-time, the system remains stable.

index 4 is the preferred one. This is due to its overall optimality compared to the rest of the actuator modes. Fig. 3.2 shows the convergence of the states under actuator MTD. Under the appropriate dwell time, the switching system remains asymptotically stable.

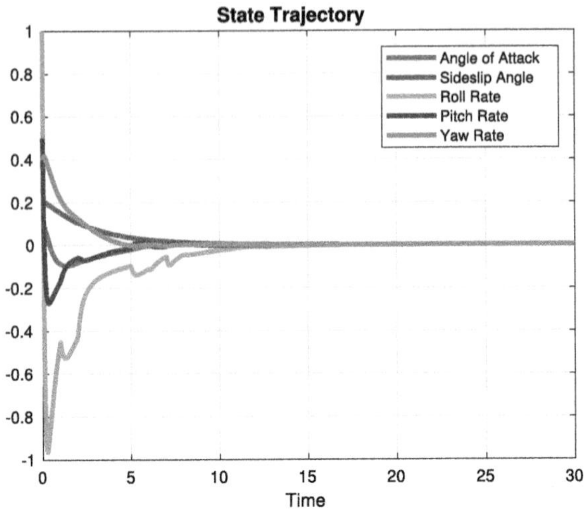

Figure 3.3 The evolution of the state with both proactive and reactive defenses. Even in the presence of attacks, the system converges to the origin since the attacked components have been taken offline.

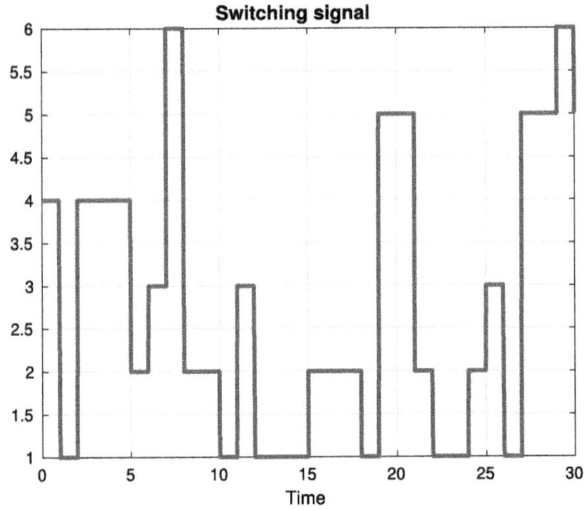

Figure 3.4 The evolution of the switching signal with both proactive and reactive defenses. We can see that when the adversarial signal is detected in the fourth controller, the random switching persists but never chooses the compromised configuration.

In Fig. 3.3, we combine the reactive and proactive security systems. The adversary manages to completely shut down one of the actuators belonging to the most optimal controller in $t = 6$ sec. It is clear that the system is stabilized. In Fig. 3.4, we show the evolution of the random switching signal favoring the controller with the best performance. After an attack is detected, the compromised component is taken out of the switching queue. However, even without the compromised mode, the MTD structure still operates. This way, we maintain some level of unpredictability while guaranteeing attack-free operation of the system. We note that the more compromised modes we have, the less unpredictable the system will be.

3.2. Graph-theoretic security index for CPS

Actuators and sensors are vital components for CPS. Their locations and numbers directly affect the control policies made by the system operators, who need to consider carefully where to put the available actuators and sensors to ensure systems operate in a desired and reliable way. On the one hand, actuators and sensors can be expensive. Thus it is important to figure out how many of them are needed in practice to be cost-efficient. On the other hand, CPS have gradually become large-scale and decentralized in recent years and rely more and more on communication networks. This high-dimensional and decentralized structure increases the exposure to malicious attacks that can cause faults, failures, and even significant damage. Therefore a security measurement is needed to guide system operators to select the numbers or locations of actuators and sensors.

In this section, we consider a discrete-time LTI system under both actuator and sensor attacks. A security index and its computation are derived based on undetectability.

3.2.1 Problem formulation

Consider the following discrete-time LTI system:

$$\begin{aligned} x_{k+1} &= Ax_k + Bu_k + B_a a_k, \\ y_k &= Cx_k + D_a a_k, \end{aligned}$$

where $k \in \mathbb{N}$ is the discrete time index, $x_k \in \mathbb{R}^n$, $u_k \in \mathbb{R}^m$, and $y_k \in \mathbb{R}^l$ are the state vector, control input, and potentially compromised output, respectively, $A \in \mathbb{R}^{n\times n}$, $B \in \mathbb{R}^{n\times m}$, and $C \in \mathbb{R}^{l\times n}$ are the state, input, and output matrices, respectively. The attack vector $a_k \in \mathbb{R}^{m+l}$ stands for the additive adversaries with the first m entries of a_k corresponding to actuator attacks, whereas the remaining l entries correspond to sensor attacks. The actuator attacks corrupt the controller command u_k by adding a value

$Ba_k(1:m,:)$ that happens during the communication from controllers to actuators, where $a_k(1:m,:)$ stands for the first m entries of a_k. Similarly, the sensor attacks replace the true measurement signals Cx_k with a corrupted value that happens during the communication from sensors to controllers. The matrices $B_a \in \mathbb{R}^{n\times(m+l)}$ and $D_a \in \mathbb{R}^{l\times(m+l)}$ represent attacker's capabilities to corrupt actuators and sensors, respectively, given by $B_a = [B \quad \mathbf{0}_{n\times l}]$ and $D_a = [\mathbf{0}_{l\times m} \quad \mathbf{I}_l]$, where $\mathbf{I}_l$ denotes an identity matrix of dimension l. The ith column vector of B corresponds to the ith actuator, whereas the jth row vector of C corresponds to the jth sensor. Let U and S denote the sets of actuators and sensors, respectively. Let $U_a \subseteq U$ be the set of attacked actuators with $|U_a| = m'$, and let $S_a \subseteq S$ be the set of attacked sensors with $|S_a| = l'$. Assume that there are no attack signals added to the safe actuators and sensors. The attacker has full information of the system dynamics, i.e., matrices A, B, and C, whereas the injected attack signals a_k are unknown to the system operator. Throughout the work, the attack signal a_k is assumed to be nonzero. The attacked actuators and sensors are assumed to be fixed, but the values of attack signals may change over time. The matrix B is assumed to have a full column rank. The pairs (A, B) and (A, C) are assumed to be controllable and observable, respectively. Generally, since the control input u_k is given by the system operator, its contribution to output can be calculated accurately and does not affect the results in this work. Due to superposition properties of LTI systems, without loss of generality, we neglect the control input term Bu_k throughout this work. Instead, we will focus on the following system, denoted as $\Sigma = (A, B, C, B_a, D_a)$: for $k \in \mathbb{N}$,

$$x_{k+1} = Ax_k + B_a a_k, \tag{3.23}$$

$$y_k = Cx_k + D_a a_k. \tag{3.24}$$

Assumption 3.5. The pairs (A, B) and (A, C) are controllable and observable, respectively. ■

3.2.2 Conditions for undetectable attacks

The security level of CPS can be measured by their ability to detect attacks. So in this section, we study undetectable attacks. During the time $0, 1, \ldots, N$ with $N \in \mathbb{Z}_{>0}$, for the system $\Sigma = (A, B, C, B_a, D_a)$, denote the corresponding output trajectory as $Y_N = [y_0^{\mathrm{T}} \quad y_1^{\mathrm{T}} \quad \ldots \quad y_N^{\mathrm{T}}]^{\mathrm{T}}$ and the corresponding unknown attack sequence as $E_N = [a_0^{\mathrm{T}} \quad a_1^{\mathrm{T}} \quad \ldots \quad a_N^{\mathrm{T}}]^{\mathrm{T}}$. The output trajectory Y_N during the time $0, \ldots, N$ is determined by the initial state x_0 and the unknown attack sequence E_N, formulated by $Y_N = O_N x_0 + V_N E_N + (\mathbf{I}_{N+1} \otimes D_a)E_N$, where $\otimes$ stands for the Kronecker product, $O_N = \begin{bmatrix} C^{\mathrm{T}} & (CA)^{\mathrm{T}} & (CA^2)^{\mathrm{T}} & \dots & (CA^N)^{\mathrm{T}} \end{bmatrix}^{\mathrm{T}}$ is the extended observability matrix, and V_N is given as

$$\begin{bmatrix} 0 & 0 & 0 & \dots & 0 & 0 \\ CB_a & 0 & 0 & \dots & 0 & 0 \\ CAB_a & CB_a & 0 & \dots & 0 & 0 \\ CA^2B_a & CAB_a & CB_a & \dots & 0 & 0 \\ \vdots & \vdots & \vdots & \vdots & \ddots & \vdots \\ CA^{N-1}B_a & CA^{N-2}B_a & CA^{N-3}B_a & \dots & CB_a & 0 \end{bmatrix}.$$

Likewise, the state vector at the time instant N is

$$x_N = A^N x_0 + C_{N-1}E_{N-1}, \tag{3.25}$$

where $C_{N-1} = \begin{bmatrix} A^{N-1}B_a & A^{N-2}B_a & \dots & B_a \end{bmatrix}$ is the extended controllability matrix. Now we introduce the following definitions.

Definition 3.2 (Dynamically Undetectable Attacks). For the system $\Sigma = (A, B, C, B_a, D_a)$, there exist dynamically undetectable attacks if and only if the nonzero attack sequence E_N satisfies $O_N x_0 + V_N E_N + (\mathbf{I}_{N+1} \otimes D_a)E_N = O_N x_0'$, $N \in \mathbb{Z}_{>0}$, with initial state x_0 and $x_0' \in \mathbb{R}^n \setminus \mathbf{0}$. ■

Next, we will provide the conditions for the existence of dynamically undetectable attacks. First, we show that it is sufficient to focus on the time period of $0, 1, \dots, n-1$ to decide whether there exists a dynamically undetectable attack sequence for the time period of $0, 1, \dots, N$, where $N \in \mathbb{Z}_{>0}$.

Lemma 3.1. *For a system $\Sigma = (A, B, C, B_a, D_a)$, assume that Assumption* 3.5 *holds and that there exists a dynamically undetectable attack sequence E_{n-1} in the time period of $0, 1, \dots, n-1$. Then there exists a dynamically undetectable attack sequence in the time period of $0, 1, \dots, N$, where $N \in \mathbb{Z}_{>0}$.*

Proof. Let there exist a dynamically undetectable attack sequence E_{n-1} during the time period of $0, 1, \dots, n-1$, i.e., $O_{n-1}x_0 + V_{n-1}E_{n-1} + (\mathbf{I}_n \otimes D_a)E_{n-1} = O_{n-1}x_0'$. Then left multiply by O_{n-1}^{T} both sides and rearrange to get $O_{n-1}^{\mathrm{T}}[V_{n-1}E_{n-1} + (\mathbf{I}_n \otimes D_a)E_{n-1}] = O_{n-1}^{\mathrm{T}}O_{n-1}(x_0' - x_0)$. The observability matrix O_{n-1} has full column rank. Then it follows that the square matrix $O_{n-1}^{\mathrm{T}} O_{n-1}$ has full rank and thus is invertible. Then $\Delta_x = x'_0 - x_0$ can be uniquely solved by $\Delta_x = (O_{n-1}^{\mathrm{T}} O_{n-1})^{-1} O_{n-1}^{\mathrm{T}} [V_{n-1} E_{n-1} + (\mathbf{I}_n \otimes D_a) E_{n-1}]$. Now consider the time instant $N = n$. Assume that there exists an attack signal a_n such that $O_n x_0 + V_n E_n + (\mathbf{I}_{n+1} \otimes D_a) E_n = O_n x'_0$ with $E_n = [E_{n-1}^{\mathrm{T}} \; a_n]^{\mathrm{T}}$. Given now (3.24) and (3.25), it follows that $y_n = CA^n x_0 + CA^{n-1} B_a a_0 + CA^{n-2} B_a a_1 + \cdots + CB_a a_{n-1} + D_a a_n = CA^n x'_0$. Rearrange to get

$$D_a a_n = CA^n \Delta_x - (CA^{n-1} B_a a_0 + \cdots + CB_a a_{n-1}). \tag{3.26}$$

Since Δ_x is uniquely solved, the right-hand-side (RHS) of (3.26) is uniquely determined. Considering $D_a = [\mathbf{0}_{l \times m} \; \mathbf{I}_l]$, the first m entries of a_n can be any values, whereas the remaining l entries of a_n are uniquely determined from (3.26). Thus the existence of a_n is guaranteed. Similarly, for $N = n+1, n+2, \ldots$, the newly added attack signal a_N always exists. □

Definition 3.3 (Perfectly Undetectable Attacks). For a system $\Sigma = (A, B, C, B_a, D_a)$, there exist perfectly undetectable attacks if and only if for a nonzero attack sequence E_N, $V_N E_N + (\mathbf{I}_{N+1} \otimes D_a) E_N = \mathbf{0}$ for all $N \in \mathbb{Z}_{>0}$. ■

Perfectly undetectable attacks leave zero trace in the sensory output. Therefore Definition 3.3 is a stricter version of Definition 3.2. Consequently, Lemma 3.1 also applies to attacks of Definition 3.3. Now we have the following corollary.

Corollary 3.1. *For a system $\Sigma = (A, B, C, B_a, D_a)$, suppose that there exists a perfectly undetectable attack sequence E_{n-1} during the time period of $0, 1, \ldots, n-1$. Then for each $N \in \mathbb{Z}_{>0}$, there exists a perfectly undetectable attack sequence during the time period of $0, 1, \ldots, N$.*

Proof. The proof follows the logic of Lemma 3.1 with $x_0 = \mathbf{0}$ and $x'_0 = \mathbf{0}$. □

Now we can only consider the time period of $0, 1, \ldots, n-1$. First, we recall the following definitions (Molinari, 1976; Trentelman et al., 2012).

Definition 3.4 (Input Unobservable Subspace). For a system $\Sigma = (A, B, C, B_a, D_a)$, the input unobservable subspace over k steps is defined as $\mathcal{I}_k = \{x \in \mathbb{R}^n$: there exists an attack sequence E_{k-1} such that $O_{k-1} x + V_{k-1} E_{k-1} + (\mathbf{I}_k \otimes D_a) E_{k-1} = \mathbf{0}\}$. ■

Definition 3.5 (Weakly Unobservable Subspace). For a system $\Sigma = (A, B, C, B_a, D_a)$, the weakly unobservable subspace, denoted as $\mathcal{W}(\Sigma)$, is defined as its input unobservable subspace over n steps, i.e., $\mathcal{W}(\Sigma) = \mathcal{I}_n$. ■

Definition 3.6 (Strongly Observable). A system $\Sigma = (A, B, C, B_a, D_a)$ is strongly observable if and only if its corresponding weakly unobservable subspace is trivial. ■

Theorem 3.3. *For a system $\Sigma = (A, B, C, B_a, D_a)$, there exists a perfectly undetectable attack sequence E_{n-1} during the time period of $0, 1, \ldots, n-1$ if and only if the system Σ is strongly observable.*

Proof. ($\Leftarrow$) Assume that system Σ is strongly observable, which implies $V_{n-1}E_{n-1} + (\mathbf{I}_n \otimes D_a)E_{n-1} = \mathbf{0}$. By Definition 3.3 this shows that there exists a perfectly undetectable attack sequence E_{n-1} for the system $\Sigma = (A, B, C, B_a, D_a)$ with initial condition $x_0 = \mathbf{0}$.

($\Rightarrow$) Assume that there exists a perfectly undetectable attack sequence E_{n-1} for the system Σ. By Definition 3.3, $V_{n-1}E_{n-1} + (\mathbf{I}_n \otimes D_a)E_{n-1} = \mathbf{0}$. If there exists $\delta \neq \mathbf{0}$ such that $O_{n-1}\delta + V_{n-1}E_{n-1} + (\mathbf{I}_n \otimes D_a)E_{n-1} = \mathbf{0}$, then $O_{n-1}\delta = \mathbf{0}$ implies that O_{n-1} does not have full column rank (contradicts Assumption 3.5). So the system Σ is strongly observable. □

Remark 3.7. The assumption that the system is strongly observable is a sufficient and necessary condition for the existence of perfectly undetectable attacks. ■

Theorem 3.4. *If a system $\Sigma = (A, B, C, B_a, D_a)$ is not strongly observable, then there exists a dynamically undetectable attack sequence E_{n-1} in the time period of $0, 1, \ldots, n-1$.*

Proof. Assume that the system Σ is not strongly observable. By Definition 3.6 there exist a nonzero $\delta \in \mathcal{W}(\Sigma)$ and an attack sequence E_{n-1} such that $O_{n-1}\delta + V_{n-1}E_{n-1} + (\mathbf{I}_n \otimes D_a)E_{n-1} = \mathbf{0}$. Define $x_0' = x_0 - \delta$ and substitute it into the above equation to get $O_{n-1}x_0 + V_{n-1}E_{n-1} + (\mathbf{I}_n \otimes D_a)E_{n-1} = O_{n-1}x_0'$. Therefore, according to Definition 3.2, there exists a dynamically undetectable attack sequence E_{n-1} during the time period of $0, 1, \ldots, n-1$. □

Corollary 3.2. *For a system $\Sigma = (A, B, C, B_a, D_a)$, the non-existence of dynamically undetectable attacks implies the existence of perfectly undetectable attacks.*

Proof. By Theorem 3.4 the non-existence of dynamically undetectable attacks implies that the system is strongly observable. By Theorem 3.3 there exist perfectly undetectable attacks. □

Note that the non-existence of dynamically undetectable attacks implies the existence of perfectly undetectable attacks. However, the existence of dynamically undetectable attacks rules out the existence of perfectly undetectable attacks. Therefore it is safe to consider perfectly undetectable attacks for system security.

3.2.3 Structured model and graph representation

The structured matrices $[A]$, $[C]$, $[B_a]$, $[D_a]$ have binary elements. The (i, j) entry of matrix $[A]$ equal to 0 means that $A_{ij} = 0$ for every realization of a matrix A, whereas $[A]_{ij} = 1$ means that A_{ij} is a free parameter and can be any value from $\mathbb{R}$ except 0. The same holds for matrices $[C]$, $[B_a]$, and $[D_a]$. Denote by $\mathcal{R}$ the set of all system realizations from structured matrices $[A]$, $[C]$, $[B_a]$, $[D_a]$. Structured systems provide less knowledge of system dynamics, but the analysis based on structured models is robust to system dynamics variations and applied to any realizations. To simplify the analysis, we make the following assumption.

Assumption 3.6. Each state is influenced directly by only one actuator and measured directly by only one sensor. ■

Assumption 3.7. Each attack signal only corrupts one actuator. ■

Remark 3.8. The formulation of the attack matrix D_a implies that each attack signal only corrupts one sensor. ■

We now associate a directed graph $\mathcal{G} = (\mathcal{V}, \mathcal{E})$ with the structured model $[A]$, $[C]$, $[B_a]$, $[D_a]$ of system (3.23)–(3.24). Denote by $\mathcal{X} = \{x^{(1)}, x^{(2)}, \ldots, x^{(n)}\}$ the set of state vertices, by $\mathcal{Y} = \{y^{(1)}, y^{(2)}, \ldots, y^{(l)}\}$ the set of output vertices, and by $\mathcal{A} = \{a^{(1)}, a^{(2)}, \ldots, a^{(m+l)}\}$ the set of attack vertices. The vertex set of $\mathcal{G}$ is formed by $\mathcal{V} = \mathcal{X} \bigcup \mathcal{Y} \bigcup \mathcal{A}$. The edge set is formed by $\mathcal{E} = \mathcal{E}_A \bigcup \mathcal{E}_C \bigcup \mathcal{E}_{B_a} \bigcup \mathcal{E}_{D_a}$, where $\mathcal{E}_A = \{(x^{(j)}, x^{(i)}) : [A]_{ij} = 1\}$ is the set of edges from vertex $x^{(j)}$ to vertex $x^{(i)}$, $\mathcal{E}_C = \{(x^{(j)}, y^{(i)}) : [C]_{ij} = 1\}$ is the set of edges from vertex $x^{(j)}$ to vertex $y^{(i)}$, $\mathcal{E}_{B_a} = \{(a^{(j)}, x^{(i)}) : [B_a]_{ij} = 1\}$ is the set of edges from vertex $a^{(j)}$ to vertex $x^{(i)}$, and $\mathcal{E}_{D_a} = \{(a^{(j)}, y^{(i)}) : [D_a]_{ij} = 1\}$ is the set of edges from vertex $a^{(j)}$ to vertex $y^{(i)}$.

Malicious attacks against the system Σ can be considered as the attacker injecting signals into the system through attack vertices in $\mathcal{A}$. Perfect undetectability means that the injected signals do not flow to output vertices in $\mathcal{Y}$. Thus we consider this problem as a flow network problem with source vertices in $\mathcal{A}$ and sink vertices in $\mathcal{Y}$. First, we add a dummy vertex t and edges from all sink/output vertices in $\mathcal{Y}$ to t. Let $\mathcal{E}_{yt} = \{(y^{(i)}, t) : \forall y^{(i)} \in \mathcal{Y},\ i = \{1, 2, \ldots, l\}\}$ denote the set of edges from vertex $y^{(i)}$ to vertex t. The vertex t is considered as an operator who receives all the measurements in the process. For flow networks, to stay perfectly undetectable, the attacker needs to prevent the flow from reaching the operator vertex t, i.e., the sensory output always equals zero. If the maximum flow from attack vertices in $\mathcal{A}$ to t is zero, then attacks remain undetectable. Let $\mathcal{G}_t = (\mathcal{V}_t, \mathcal{E}_t)$ be the extended graph of the system with the above modifications, i.e., $\mathcal{V}_t = \mathcal{V} \bigcup t$ and $\mathcal{E}_t = \mathcal{E} \bigcup \mathcal{E}_{yt}$.

3.2.4 Characterization of security index

Based on the previous discussions on undetectable attacks, we define the security index in terms of perfect undetectability.

Definition 3.7 (Security Index). For a system $\Sigma = (A, B, C, B_a, D_a)$, the security index is defined as the minimal number of attacked sensors and actuators to conduct perfectly undetectable attacks:

$$s_0 = \min_{a_k} ||a_k||_0 \tag{3.27}$$

$$\text{s.t.} \quad x_{k+1} = Ax_k + B_a a_k, \tag{3.28}$$

$$\mathbf{0} = Cx_k + D_a a_k, \tag{3.29}$$

$$x_0 = \mathbf{0}, \tag{3.30}$$

where $||a_k||_0 = |\text{supp}(a_k)|$ and $\text{supp}(a_k) = \{i \in \mathcal{I} : a_k^{(i)} \neq 0\}$, with nonzero $a_k^{(i)}$ being the ith element of a_k and $\mathcal{I}$ being the set of indices of elements of a_k. ■

Constraints (3.28) and (3.29) ensure that the system dynamics are obeyed. Constraints (3.29) and (3.30) imply that perfectly undetectable attacks are considered. Note that if (3.27) has no solution, which implies that the system Σ is not strongly observable based on Theorem 3.3, then the security index $s_0 = \infty$. If $s_0 = m + l$, then the system Σ is maximally secure, which implies that adversaries have to attack all the available actuators and sensors to remain perfectly undetectable. The computation of security index is generally NP-hard due to l_0 norm in the objective function (Natarajan, 1995). As a result, its computation is not efficient for high-dimensional systems. Next, we will rely on structured models of systems to compute the security index.

Due to Assumption 3.6 and Remark 3.8, we denote by $\mathcal{A}_a$ the set of attack vertices corresponding to the set of attacked actuators U_a and by $\mathcal{A}_s$ the set of attack vertices corresponding to the set of attacked sensors S_a. It follows that $\mathcal{A}_a \bigcup \mathcal{A}_s \subseteq \mathcal{A}$, $|\mathcal{A}_a| = m'$, and $|\mathcal{A}_s| = l'$.

Definition 3.8 (Vertex Separator). For a directed graph $\mathcal{G} = (\mathcal{V}, \mathcal{E})$, a vertex separator for nonadjacent vertices s and t is a subset of vertices in $\mathcal{V} \setminus \{s, t\}$ whose removal eliminates all the directed paths from s to t. ■

Theorem 3.5. *Consider the extended graph $\mathcal{G}_t = (\mathcal{V}_t, \mathcal{E}_t)$. For each vertex $a_k^{(i)} \in \mathcal{A}_a$, $i \in \{1, 2, \ldots, |\mathcal{A}_a|\}$, define $\mathcal{X}_a^i = \{x_k^{(j)} \in \mathcal{X} : (a_k^{(q)}, x_k^{(j)}) \in \mathcal{E}_{B_a}, a_k^{(q)} \in \mathcal{A}_a \setminus a_k^{(i)}\}$. There exist perfectly undetectable attacks for the system Σ of any realization from $\mathcal{R}$ with the set of attacked actuators U_a and the set of attacked sensor S_a if and only if $\mathcal{X}_a^i \bigcup S_a$ is a vertex separator of $a_k^{(i)}$ and t in $\mathcal{G}_t$.*

Proof. ($\Leftarrow$) Given $\mathcal{X}_a^i \bigcup S_a$ is a vertex separator of $a_k^{(i)}$ and t in $\mathcal{G}_t$, by Assumption 3.6, let the qth actuator attack signal $a_k^{(q)}$ corrupting the jth state $x_k^{(j)}$ be

$$a_k^{(q)} = -A(j, :)x_k / B_a(j, q), \quad k \in \mathbb{N}, \tag{3.31}$$

where $A(j, :)$ denotes the jth row of A, and $B_a(j, q)$ is the (j, q)th element of B_a. Note that $B_a(j, q) \neq 0$ due to Assumptions 3.6 and 3.7. For the pth attacked sensor, let the corresponding sensor attack signal be

$$a_k^{(m+p)} = -C(p, :)x_k, \quad k \in \mathbb{N}, \tag{3.32}$$

where $C(p, :)$ denotes the pth row of C. Next, we prove that attacks defined by (3.31) and (3.32) are perfectly undetectable, i.e., $y \equiv \mathbf{0}$ with $x_0 = \mathbf{0}$. For $x_k^{(j)} \in \mathcal{X}_a^i$ with $a_k^{(q)}$ influencing $x_k^{(j)}$, applying Eq. (3.31), we have $x_{k+1}^{(j)} = A(j, :)x_k + B_a(j, q)a_k^{(q)} = 0$, which implies that all states in $\mathcal{X}_a^i$ equal 0. For the pth attacked sensor, due to Eq. (3.32), we have $y_k^{(p)} = C(p, :)x_k + a_k^{(m+p)} = 0$. Next, we define $\mathcal{X}_b^i = \{x_k^{(j)} \in \mathcal{X}$: there exists a directed path from $a_k^{(i)}$ to $x_k^{(j)}$ that does not include states in $\mathcal{X}_a^i\}$. We claim that states in $\mathcal{X}_b^i$ cannot be measured by attack-free sensors. If so, then there exists a directed path from $a_k^{(i)}$ to t, which contradicts that $\mathcal{X}_a^i \bigcup \mathcal{A}_s$ is a vertex separator of $a_k^{(i)}$ and t in $\mathcal{G}_t$. For the remaining states $\mathcal{X}_c^i = \mathcal{X} \setminus (\mathcal{X}_a^i \bigcup \mathcal{X}_b^i)$, we claim that an edge $(x_k^{(b)}, x_k^{(c)})$ with $x_k^{(b)} \in \mathcal{X}_b^i$ and $x_k^{(c)} \in \mathcal{X}_c^i$ does not exist. If so, then this would imply that there exists a directed path from $a_k^{(i)}$ to $x_k^{(c)}$ that does not include states in $\mathcal{X}_a^i$. Then by the definition of $\mathcal{X}_b^i$ we get $x_k^{(c)} \in \mathcal{X}_b^i$, which is a contradiction since $x_k^{(c)} \in \mathcal{X}_c^i$. Thus we conclude that states in $\mathcal{X}_c^i$ are not affected by states in $\mathcal{X}_b^i$. Since $x_0 = \mathbf{0}$ and we have proved that states in $\mathcal{X}_a^i$ equal 0, then states in $\mathcal{X}_c^i$ always remain 0. We have showed that states in $\mathcal{X}_b^i$ cannot be measured by attack-free sensors. Thus the attack-free sensor measurement equals 0. We have proved that the attacked sensor measurement remains 0. Therefore attacks with strategies of (3.31) and (3.32) are perfectly undetectable.

($\Rightarrow$) Suppose that $\mathcal{X}_a^i \bigcup \mathcal{S}_a$ is not a vertex separator of $a_k^{(i)}$ and t. Then it follows that there exists a directed path from $a_k^{(i)}$ to t that does not include states in $\mathcal{X}_a^i \bigcup \mathcal{A}_s$. We denote this path as $p_i = \{a_k^{(i)} \to x_k^{(i_1)} \to x_k^{(i_2)} \to \cdots \to x_k^{(i_n)} \to y_k^{(q)} \to t\}$. Now we need to show that no perfectly undetectable attacks for at least one realization from $\mathcal{R}$ exist. For $x_k^{(i_1)}$ from path p_i, let $A(i_1, :) = 0$, so that other states cannot affect $x_k^{(i_1)}$. For other states $x_k^{(i_j)}$ from path p_i, where $2 \le j \le n$, set $A(i_j, h) \neq 0$ for $h = i_{j-1}$ and $A(i_j, h) = 0$ for $h \neq i_{j-1}$, so that only $x_k^{(i_{j-1})}$ can affect $x_k^{(i_j)}$. Let $C(q, i_n) \neq 0$, so that $y_k^{(q)} \neq 0$ as long as $x_k^{(i_n)} \neq 0$ due to Assumption 3.6. For $a_k^{(i)} \neq 0$, given $A(i_1, :) = 0$, we have $x_{k+1}^{(i_1)} = B(i_1, i)a_k^{(i)} \neq 0$. Now for other states from path p_i, we have $x_{k+2}^{(i_2)} = A(i_2, i_1)x_{k+1}^{(i_1)} \neq 0 \Rightarrow x_{k+3}^{(i_3)} = A(i_3, i_2)x_{k+2}^{(i_2)} \neq 0 \Rightarrow \dots \Rightarrow x_{k+n}^{(i_n)} = A(i_n, i_{n-1})x_{k+n-1}^{(i_{n-1})} \neq 0$. Since $C(q, i_n) \neq 0$, we have $y_{k+n+1}^{(q)} = C(q, i_n)x_{k+n}^{(i_n)} \neq 0$. Therefore there does not exist perfectly undetectable attacks for this realization. □

Theorem 3.5 formulates the condition for the existence of perfectly undetectable attacks from the perspective of actuator attack signal $a_k^{(i)}$, $i \in \{1, 2, \dots, |\mathcal{A}_a|\}$. Note that for attacks to remain perfectly undetectable, at least one actuator needs to be attacked to hide sensor attack signals. Thus we have the following corollary to formulate the condition for the existence of perfectly undetectable attacks from the view of sensor attack signals.

Corollary 3.3. *Consider the extended graph $\mathcal{G}_t = (\mathcal{V}_t, \mathcal{E}_t)$. For each vertex $a_k^{(j)} \in \mathcal{A}_s$ with $j \in \{1, 2, \dots, |\mathcal{A}_s|\}$, suppose the corresponding jth attacked sensor measures the pth state $x_k^{(p)}$. Let $a_k^{(i)}$ denote the actuator attack signal that attacks the pth state directly (i.e., there exists an edge from $a_k^{(i)}$ to $x_k^{(p)}$) or indirectly (i.e., there exists a directed path from $a_k^{(i)}$ to $x_k^{(p)}$). Define $\mathcal{X}_a^i = \{x_k^{(j)} \in \mathcal{X} : (a_k^{(q)}, x_k^{(j)}) \in \mathcal{E}_{B_a}, a_k^{(q)} \in \mathcal{A}_a \setminus a_k^{(i)}\}$. Then there exist perfectly undetectable attacks for the system Σ of any realization from $\mathcal{R}$ with the set of attacked actuators U_a and the set of attacked sensor S_a if and only if $\mathcal{X}_a^i \bigcup S_a$ is a vertex separator of $a_k^{(i)}$ and t in $\mathcal{G}_t$.*

Proof. Due to Assumption 2.4, Assumption 3.1, and Remark 3.1, for the jth attacked sensor measuring the pth state with attack signal $a_k^{(j)} \in \mathcal{A}_s$, for the existence of perfectly undetectable attacks, i.e., $y \equiv \mathbf{0}$ with $x_0 = \mathbf{0}$, there must exist an actuator attack signal denoted as $a_k^{(i)} \in \mathcal{A}_a$ directly or indirectly corrupting the pth state. Thus, following the logic of the proof of Theorem 3.5, $\mathcal{X}_a^i \bigcup \mathcal{S}_a$ being a vertex separator of $a_k^{(i)}$ and t with $y_k^{(j)} \in \mathcal{S}_a$ is sufficient and necessary conditions for the existence of perfectly undetectable attacks. □

Remark 3.9. Compared to Theorem 3.5, Corollary 3.3 requires the nonzero sensor attack signal $a_k^{(j)}$. The condition for the existence of perfectly undetectable attacks from the perspective of sensor attacks can be reduced to that from the perspective of actuator attacks. ■

Definition of security index (3.27)–(3.30) aims to find the minimum number of attacked actuators and sensors for adversaries to remain perfectly undetectable. Theorem 3.5 and Corollary 3.3 characterize the conditions for the existence of perfectly undetectable attacks. Therefore we can compute the security index by solving a problem of finding the minimum size of $\mathcal{A}_a \bigcup \mathcal{A}_s$ such that $\mathcal{X}_a^i \bigcup \mathcal{S}_a$ with $i \in \{1, 2, \ldots, |\mathcal{A}_a|\}$ is a vertex separator of $a_k^{(i)}$ and t in $\mathcal{G}_t$.

Problem 3.1. For a system $\Sigma = (A, B, C, B_a, D_a)$, the security index s is computed as

$$
\begin{aligned}
& s = \min_{\mathcal{A}_a, \mathcal{A}_s} (|\mathcal{A}_a \bigcup \mathcal{A}_s|) \\
\text{s.t.} \quad & \forall i \in \{1, 2, \ldots, |\mathcal{A}_a|\}, \\
& \mathcal{X}_a^i \bigcup \mathcal{S}_a \text{ is a vertex separator of } a_k^{(i)} \text{ and } t \text{ in } \mathcal{G}_t \text{ with} \\
& \mathcal{X}_a^i = \{x_k^{(j)} \in \mathcal{X} : (a_k^{(q)}, x_k^{(j)}) \in \mathcal{E}_{B_a}, a_k^{(q)} \in \mathcal{A}_a \setminus a_k^{(i)}\}.
\end{aligned}
$$

3.2.5 Computation of security index

Now we focus on explicitly characterizing the value of security index. As we discussed above, the extended graph $\mathcal{G}_t$ can be considered as flows represented by attack signals from source vertices in $\mathcal{A}_a \bigcup \mathcal{A}_s$ to the system operator t. For attacks to remain perfectly undetectable, they need to prevent the flow from reaching t. Inspired by Milošević et al. (2018), now we convert the extended graph $\mathcal{G}_t = (\mathcal{V}_t, \mathcal{E}_t)$ to a flow network $\mathcal{G}' = (\mathcal{V}', \mathcal{E}')$ by adding a flow capacity for each edge. For each $i \in \{1, 2, \ldots, |\mathcal{A}_a|\}$, create $\mathcal{V}'$ and $\mathcal{E}'$ as follows.

Rule 1. For each vertex $x_k^{(j)} \in \mathcal{X}_a^i$, split $x_k^{(j)}$ into two vertices $x_{k,1}^{(j)}$ and $x_{k,2}^{(j)}$ with an edge from $x_{k,1}^{(j)}$ to $x_{k,2}^{(j)}$ and a flow capacity the same to the incoming flow to $x_{k,1}^{(j)}$.

Rule 2. For each vertex $x_k^{(q)} \in \mathcal{X} \setminus \mathcal{X}_a^i$, keep $x_k^{(q)}$ in $\mathcal{G}'$.

Rule 3. Keep $(a_k^{(j)}, x_k^{(q)}) \in \mathcal{E}_{B_a}$ in $\mathcal{G}'$ with flow capacity ∞.

Rule 4. Consider $(x_k^{(j)}, x_k^{(q)}) \in \mathcal{E}_A$.

- For $x_k^{(j)} \in \mathcal{X}_a$ and $x_k^{(q)} \in \mathcal{X}_a$, include $(x_{k,2}^{(j)}, x_{k,1}^{(q)})$ in $\mathcal{E}'$ with flow capacity ∞.
- For $x_k^{(j)} \in \mathcal{X}_a^i$ and $x_k^{(q)} \in \mathcal{X} \setminus \mathcal{X}_a^i$, include $(x_{k,2}^{(j)}, x_k^{(q)})$ in $\mathcal{E}'$ with flow capacity ∞.
- For $x_k^{(j)} \in \mathcal{X} \setminus \mathcal{X}_a^i$ and $x_k^{(q)} \in \mathcal{X}_a^i$, include $(x_k^{(j)}, x_{k,1}^{(q)})$ in $\mathcal{E}'$ with flow capacity ∞.
- For $x_k^{(j)} \in \mathcal{X} \setminus \mathcal{X}_a^i$ and $x_k^{(q)} \in \mathcal{X} \setminus \mathcal{X}_a^i$, include $(x_k^{(j)}, x_k^{(q)})$ in $\mathcal{E}'$ with flow capacity ∞.

Rule 5. For vertex $x_k^{(j)} \in \mathcal{X}_a^i$, include $(x_{k,1}^{(j)}, x_{k,2}^{(j)})$ in $\mathcal{E}'$ with a flow capacity 1.

Rule 6. Consider $(x_k^{(j)}, y_k^{(q)}) \in \mathcal{E}_C$. For $x_k^{(j)} \in \mathcal{X}_a^i$, include $(x_{k,2}^{(j)}, y_k^{(q)})$ in $\mathcal{E}'$ with flow capacity 1. For $x_k^{(j)} \in \mathcal{X} \setminus \mathcal{X}_a^i$, include $(x_k^{(j)}, y_k^{(q)})$ in $\mathcal{E}'$ with flow capacity 1.

Rule 7. Consider $(y_k^{(j)}, t)$. If $y_k^{(j)}$ is not attacked, then include $(y_k^{(j)}, t)$ in $\mathcal{E}'$ with flow capacity ∞. If $y_k^{(j)}$ is attacked, then include $(y_k^{(j)}, t)$ in $\mathcal{E}'$ with flow capacity 1.

Given Assumptions 3.6 and 3.7 and Remark 3.8, when assigning edge flow capacities, Rule 5 guarantees that each attack signal only corrupts one actuator and each actuator directly influences on state. Rule 6 guarantees that each sensor directly measures one state. Rule 7 guarantees that each attack signal only corrupts one sensor. Next, we derive a relationship between the minimum size of vertex separator and minimum cut.

Definition 3.9 ($s-t$ cut and its capacity). For a directed graph $\mathcal{G} = (\mathcal{V}, \mathcal{E})$ with designated source vertex s and sink vertex t, along with a flow capacity c_{ij} for each edge $(v_i, v_j) \in \mathcal{E}$, a partition of $\mathcal{V}$ into disjoint sets L and R such that $\mathcal{V} = L \bigcup R$ as well as $s \in L$ and $t \in R$ is called an $s-t$ cut. The cut capacity is defined as $C(L, R) \triangleq \sum_{(i,j)\in\mathcal{E}|i\in L, j\in R} c_{ij}$. ■

Definition 3.10 (Minimal $s-t$ cut problem). The minimal $s-t$ cut problem refers to finding an $s-t$ cut on $\mathcal{G} = (\mathcal{V}, \mathcal{E})$ with the minimum capacity. ■

Theorem 3.6. *For the flow network $\mathcal{G}' = (\mathcal{V}', \mathcal{E}')$, the minimum size of vertex separator $\mathcal{X}_a^i \bigcup \mathcal{S}_a$ from $a_k^{(i)}$ to t is equivalent to the minimum $a_k^{(i)} - t$ cut.*

Proof. For $x_k^{(j)} \in \mathcal{X}_a^i$, $x_k^{(j)}$ being a vertex separator implies that there is a cut of the edge from $x_{k,1}^{(j)}$ to $x_{k,2}^{(j)}$ in $\mathcal{G}'$ with flow capacity 1. For $y_k^{(j)} \in \mathcal{S}_a$, $y_k^{(j)}$ being a vertex separator implies that there is a cut of the edge from $y_k^{(j)}$ to t with flow capacity 1. Note that $\mathcal{X}_a^i \bigcup \mathcal{S}_a$ being a vertex separator from $a_k^{(i)}$ to t means that the removal of all vertices in $\mathcal{X}_a^i \bigcup \mathcal{S}_a$ eliminates all the directed paths from $a_k^{(i)}$ to t, and one vertex in $\mathcal{X}_a^i \bigcup \mathcal{S}_a$ corresponds to a cut of edge with flow capacity 1. Therefore the minimum size of vertex separator $\mathcal{X}_a^i \bigcup \mathcal{S}_a$ from $a_k^{(i)}$ to t is equivalent to the minimum $a_k^{(i)} - t$ cut in $\mathcal{G}'$. □

For the flow network $\mathcal{G}' = (\mathcal{V}', \mathcal{E}')$, let δ be the minimum capacity of $a_k^{(i)} - t$ on $\mathcal{G}'$ for all $i \in \{1, 2, \dots, |\mathcal{A}_a|\}$. Then by Theorem 3.6 the solution to Problem 3.1 is $s = \delta + 1$, with 1 added due to the consideration of the attacked actuator associated with the attack signal $a_k^{(i)}$. Based on the max-flow min-cut theorem, the minimum capacity of an $s - t$ cut equals to the size of maximum flow from s to t. Finding the maximum flow on a directed graph is a standard max-flow problem, which can be solved by the Ford–Fulkerson or Edmonds–Karp algorithm in polynomial time.

3.2.6 Numerical examples

Consider the following structured system model with five states:

$$[A] = \begin{bmatrix} 1 & 0 & 0 & 1 & 1 \\ 0 & 1 & 0 & 0 & 0 \\ 0 & 0 & 1 & 0 & 1 \\ 0 & 1 & 0 & 1 & 0 \\ 0 & 0 & 1 & 0 & 1 \end{bmatrix}, [B] = \begin{bmatrix} 0 & 0 \\ 1 & 0 \\ 0 & 0 \\ 0 & 0 \\ 0 & 1 \end{bmatrix}, [C] = \begin{bmatrix} 0 & 1 & 0 & 0 & 0 \\ 0 & 0 & 1 & 0 & 0 \\ 0 & 0 & 0 & 0 & 1 \end{bmatrix}. \tag{3.33}$$

There are two actuators ($u_k^{(1)}$, $u_k^{(2)}$) and three sensors ($y_k^{(1)}$, $y_k^{(2)}$, $y_k^{(3)}$). The actuators $u_k^{(1)}$ and $u_k^{(2)}$ directly affect the states $x_k^{(2)}$ and $x_k^{(5)}$, respectively. The sensors $y_k^{(1)}$, $y_k^{(2)}$, and $y_k^{(3)}$ directly measure the states $x_k^{(2)}$, $x_k^{(3)}$, and $x_k^{(5)}$, respectively. Assume that both actuators and only the second sensor are corrupted by malicious attacks, i.e., $a_k^{(1)} \neq 0$, $a_k^{(2)} \neq 0$, $a_k^{(4)} \neq 0$. The extended graph representation $\mathcal{G}_t$ for the structured model (3.33) is shown as Fig. 3.5.

For $i = 1$, $\mathcal{X}_a^1 = \{x_k^{(5)}\}$, whereas for $i = 2$, $\mathcal{X}_a^2 = \{x_k^{(2)}\}$. The corresponding

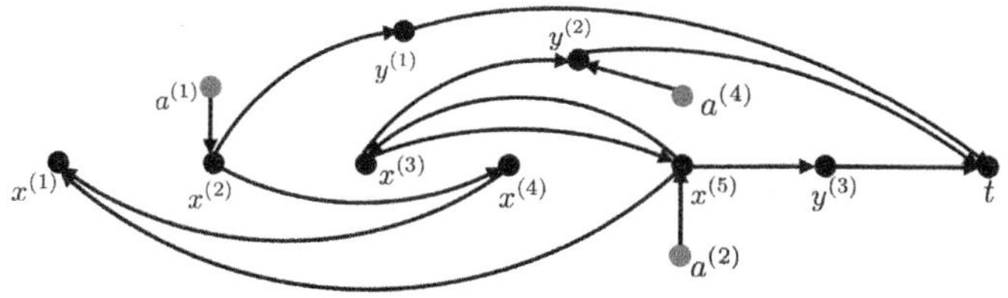

Figure 3.5 Extended graph for the structured system model (3.33).

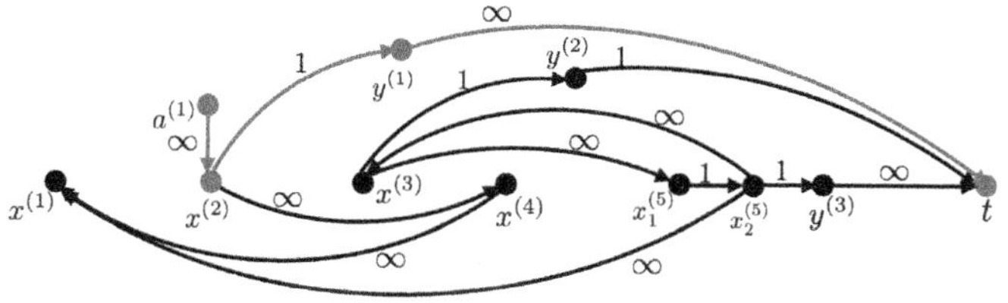

Figure 3.6 Flow network from $a^{(1)}$ to t with highlighted max-flow path.

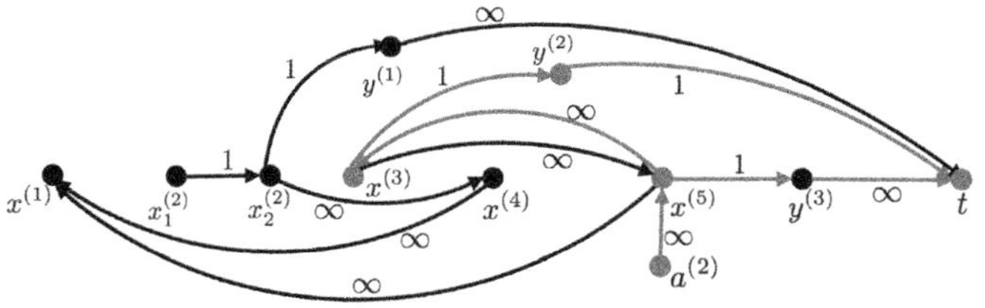

Figure 3.7 Flow network from $a^{(2)}$ to t with highlighted max-flow path.

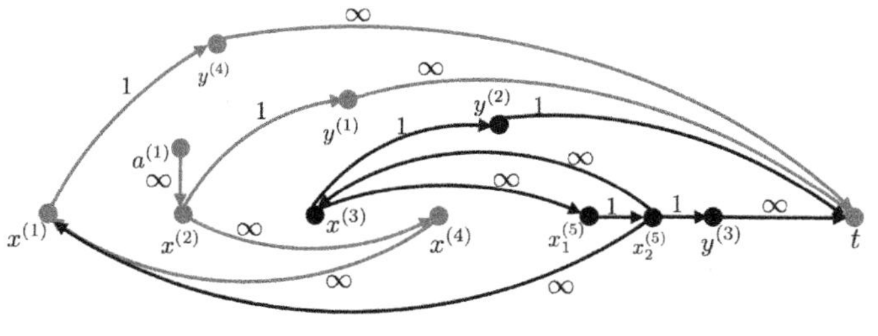

Figure 3.8 Flow network from $a^{(1)}$ to t after adding $y^{(4)}$ to measure $x^{(1)}$.

flow networks $\mathcal{G}'$ from the perspective of $a_k^{(1)}$ and $a_k^{(2)}$ are shown as Figs. 3.6 and 3.7, respectively. The security index is solved as $s = \min\{1, 2\} + 1 = 2$ with max-flow paths from a to t highlighted in red (mid gray in print version). The security index $s = 2$ implies that adversaries can only attack

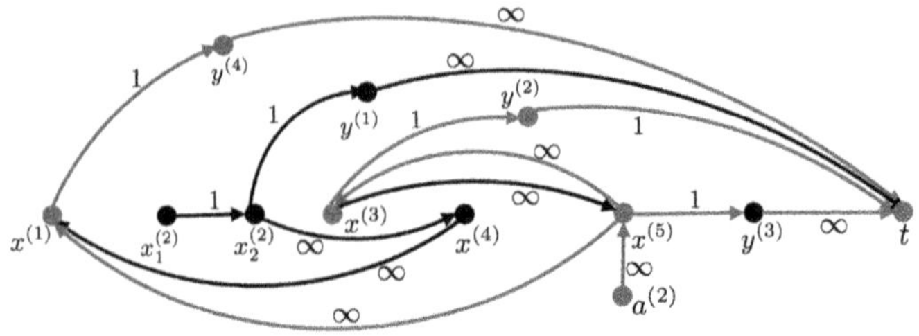

Figure 3.9 Flow network from $a^{(2)}$ to t after adding $y^{(4)}$ to measure $x^{(1)}$.

the first actuator $u_k^{(1)}$, whereas the other actuator or sensor remains undetectable. Next, we aim to increase the security index by placing more actuators or sensors. Specifically, a secure sensor $y_k^{(4)}$ is added to directly measure $x_k^{(1)}$. Now the corresponding flow networks are shown as Figs. 3.8 and 3.9 with max-flow paths from a to t highlighted in red (mid gray in print version). The security index is now $s' = \min\{2, 3\} + 1 = 3$, which shows that the placement of one more secure sensor makes the system less vulnerable.

CHAPTER FOUR

Timing faults and attacks

4.1. Timing issues in CPS

Cyber-physical systems are systems that are highly distributed, comprising multiple physical and virtual components that are interconnected through a variety of communication channels. The complexity of CPS generates several issues not usually observed in less complex systems. One relevant issue in CPS is the need to have precise timing. Generally speaking, when a control or learning algorithm is designed, it is usually assumed that all of the system components are appropriately synchronized and that their internal clocks are in agreement. Nevertheless, the distributed nature of CPS renders such a requirement difficult to guarantee, or even verify. In fact, asynchrony between the CPS component clocks, as well as timing errors, are almost always expected to occur in reality. These errors can be quite disruptive to the underlying process that controls the CPS plant, possibly leading to loss of closed-loop performance or even instability. Thus it is important to study the possible effect of timing errors when designing a particular control or learning process.

4.2. Clock offsets during autonomous decision-making for discrete-time systems

In practice, CPS are highly distributed systems with computer nodes and communication networks connecting various components. Efficient and reliable operation of CPS requires precise timing: all components share the same time at the same time instant. Thus timing is of critical importance to the flawless functionality and resilience of CPS, but clock asynchronization and time delays among components are quite ubiquitous in these distributed systems. Moreover, for large-scale CPS, accurate system models are difficult or impossible to obtain, and utilizing data generated by systems for control synthesis is a tendency to go. Clock offsets among different components may induce and propagate mismatched signals in the system and thus negatively impact performance. Therefore understanding the impact of asynchronous clocks on data-driven control is an important research issue.

Control and Game Theoretic Methods for Cyber-Physical Security
https://doi.org/10.1016/B978-0-44-315408-9.00010-5

In this section, we formulate the problem of data-driven off-policy RL with clock offsets. Impacts of clock offsets on the data-driven off-policy RL are described.

4.2.1 System setup under clock offsets

Consider a discrete-time linear time-invariant system

$$x_{k+1} = Ax_k + Bu_k, \tag{4.1}$$

where $k \in \mathbb{N}$ is the discrete time index, $x_k \in \mathbb{R}^n$ is the state vector with initial condition $x_0 \in \mathbb{R}^n$, $u_k \in \mathbb{R}^m$ is the control input, $A \in \mathbb{R}^{n\times n}$ is the state matrix, and $B \in \mathbb{R}^{n\times m}$ is the input matrix. The pair (A, B) is assumed to be stabilizable. The general structure of a CPS that incorporates learning is shown in Fig. 4.1. Each system component is spatially distributed, i.e., the actuators, the physical plant, sensors, controller, and the learning component all have their own distinct clocks, which ideally should be synchronized with one another. However, imperfect communication channels and hardware might lead to timing mismatches, which in turn can lead to inconsistent data propagation throughout the system. Such corrupted data can jeopardize the nominal function of the learning component, which inherently assumes that all clocks in the CPS agree with one another.

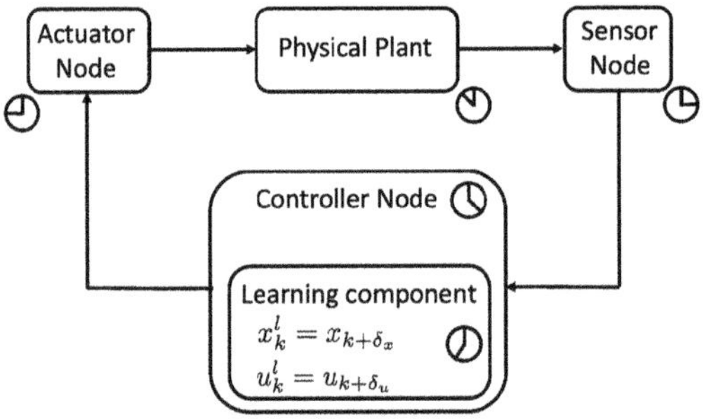

Figure 4.1 CPS structure with asynchronous clocks.

Specifically speaking, the true state and the true control input at time instant k are ideally x_k and u_k, respectively. However, owing to timing discrepancies among components of CPS, the actual state and control input signals perceived by the learning component at each time instant k are $x_k^l = x_{k+\delta_x(k)}$ and $u_k^l = u_{k+\delta_u(k)}$, respectively, where $\delta_x : \mathbb{N} \to \mathbb{Z}$ is the clock offset for state signals, and $\delta_u : \mathbb{N} \to \mathbb{Z}$ is the clock offset for control input signals. In the upcoming sections, we will study the impact of these discrepancies between the actual states/control signals and the perceived ones, particularly conditions of achieving a tolerable learning behavior under clock offsets for system (4.1).

4.2.2 Preliminaries: optimal control and model-based RL

For a given stabilizing control policy $\mu : \mathbb{R}^n \to \mathbb{R}^m$, define its corresponding performance criterion as

$$J(x_0; \mu) = \sum_{k=0}^{\infty} \left(x_k^{\mathrm{T}} Q x_k + \mu^{\mathrm{T}}(x_k) R \mu(x_k) \right), \tag{4.2}$$

where $Q \in \mathbb{R}^{n \times n} \succeq 0$ and $R \in \mathbb{R}^{m \times m} \succ 0$, and the pair $(A, \ Q)$ is observable. Then the value function V^{μ} corresponding to the policy μ, defined as $V^{\mu}(\cdot) := J(\cdot; \ \mu)$, satisfies the difference form of Bellman equation: for all $k \in \mathbb{N}$,

$$V^{\mu}(x_k) = x_k^{\mathrm{T}} Q x_k + \mu^{\mathrm{T}}(x_k) R \mu(x_k) + V^{\mu}(x_{k+1}). \tag{4.3}$$

The goal is to find a policy $\mu^\star : \mathbb{R}^n \to \mathbb{R}^m$ that minimizes (4.2) yielding the optimal value function, i.e., $V^\star(x_k) = \min_{\mu} J(x_k; \mu)$. In line with the literature on linear-quadratic control, consider the linear state feedback policy and its quadratic value function:

$$\mu(x_k) = -Kx_k, \tag{4.4}$$

$$V^{\mu}(x_k) = x_k^{\mathrm{T}} P x_k, \tag{4.5}$$

with $P \in \mathbb{R}^{n \times n} \succ 0$. Using (4.4) and (4.5), the Bellman equation (4.3) can then be simplified to the Lyapunov equation

$$(A - BK)^{\mathrm{T}} P (A - BK) + Q + K^{\mathrm{T}} R K = P. \tag{4.6}$$

Now define the Hamiltonian: for all x_k, μ, and k,

$$H(x_k, \mu) = x_k^{\mathrm{T}} Q x_k + \mu^{\mathrm{T}} R \mu + x_{k+1}^{\mathrm{T}} P x_{k+1} - x_k^{\mathrm{T}} P x_k. \tag{4.7}$$

Then the optimal control $\mu^\star(x_k) = K^\star x_k$ can be derived by applying the stationarity condition on the Hamiltonian, which yields the optimal gain

$$K^\star = (R + B^{\mathrm{T}} P^\star B)^{-1} B^{\mathrm{T}} P^\star A, \tag{4.8}$$

where $P^\star$ satisfies the algebraic Riccati equation (ARE)

$$0 = A^{\mathrm{T}} P^\star A - P^\star + Q - A^{\mathrm{T}} P^\star B (R + B^{\mathrm{T}} P^\star B)^{-1} B^{\mathrm{T}} P^\star A. \tag{4.9}$$

Note that the ARE (4.9) is a nonlinear function of P, which is difficult to solve directly, especially for large-scale systems. Various approaches have been proposed to find its solution numerically. One such iterative approach, known as the model-based policy-iteration (PI) (Hewer, 1971) is presented in Algorithm 4.1.

Algorithm 4.1 Model-Based RL.

Select a stabilizing controller gain K^0. Set $j=0$.

1. Solve for P^{j+1} such that

$$(A-BK^j)^{\mathrm{T}}P^{j+1}(A-BK^j)+Q+(K^j)^{\mathrm{T}}RK^j=P^{j+1}. \tag{4.10}$$

2. Compute the next gain as

$$K^{j+1}=(R+B^{\mathrm{T}}P^{j+1}B)^{-1}B^{\mathrm{T}}P^{j+1}A.$$

3. Stop if $\|K^{j+1}-K^j\|\le\xi$ with a positive threshold ξ. Otherwise, let $j=j+1$ and go to step 1.

4.2.3 Data-driven off-policy RL without clock offsets

Algorithm 4.1 for solving the ARE (4.9) is inherently model-based, which requires the knowledge of system matrices A and B. Nevertheless, various RL techniques are proposed to waive this knowledge requirement to solve (4.9) in a data-driven manner. One of these techniques, known as off-policy RL (Kiumarsi et al., 2017), begins by rewriting the original system (4.1) as

$$x_{k+1}=A_Kx_k+B(Kx_k+u_k) \tag{4.11}$$

with $A_K=A-BK$. Here u_k can be any policy rather than the desired optimal one, hence justifying the "off-policy" label of this algorithm. It is generally selected as $u_k=\mu_{\mathrm{b}}(x_k)+e_k$, where μ_{b} is an input-to-state stabilizing behavioral policy, and $e_k\in\mathbb{R}^m$ is an exploration noise vector added to satisfy the persistence excitation (PE) condition. This control input is then applied to systems to generate input-state data, which are used to express (4.10) in a data-based manner. Manipulating (4.5), (4.10), and (4.11) yields the off-policy Bellman equation

$$\begin{aligned}&x_k^{\mathrm{T}}P^{j+1}x_k-x_{k+1}^{\mathrm{T}}P^{j+1}x_{k+1}\\&=x_k^{\mathrm{T}}Qx_k+x_k^{\mathrm{T}}(K^j)^{\mathrm{T}}RK^jx_k-(u_k+K^jx_k)^{\mathrm{T}}B^{\mathrm{T}}P^{j+1}A_Kx_k\end{aligned}$$

$$- (u_k + K^j x_k)^{\mathrm{T}} B^{\mathrm{T}} P^{j+1} x_{k+1}. \tag{4.12}$$

Using the property of Kronecker product, $x^{\mathrm{T}} Ay = (y^{\mathrm{T}} \otimes x^{\mathrm{T}})\mathrm{vec}(A)$, and considering the reconstructed dynamics (4.11), (4.12) can be rewritten as follows:

$$\begin{aligned} 0 = & - (x_k^{\mathrm{T}} \otimes x_k^{\mathrm{T}})\mathrm{vec}(P^{j+1}) + (x_{k+1}^{\mathrm{T}} \otimes x_{k+1}^{\mathrm{T}})\mathrm{vec}(P^{j+1}) \\ & - 2(x_k^{\mathrm{T}} \otimes (u_k + K^j x_k)^{\mathrm{T}})\mathrm{vec}(B^{\mathrm{T}} P^{j+1} A) \\ & - ((u_k - K^j x_k)^{\mathrm{T}} \otimes (u_k + K^j x_k)^{\mathrm{T}})\mathrm{vec}(B^{\mathrm{T}} P^{j+1} B) \\ & + x_k^{\mathrm{T}} Q x_k + x_k^{\mathrm{T}} (K^j)^{\mathrm{T}} R K^j x_k. \end{aligned} \tag{4.13}$$

Lemma 4.1. *Consider the trajectories generated by system* (4.11). *Then the solution* P^{j+1} *of* (4.10) *is also a solution of the data-based equation* (4.13).

Proof. For the jth iteration of the learning process, it follows from the quadratic value function (4.5) and dynamics (4.11) that for all $k \in \mathbb{N}$,

$$\begin{aligned} & V^{j+1}(x_k) - V^{j+1}(x_{k+1}) \\ = & x_k^{\mathrm{T}} P^{j+1} x_k - x_{k+1}^{\mathrm{T}} P^{j+1} x_{k+1} \\ = & x_k^{\mathrm{T}} P^{j+1} x_k - [A_K x_k + B(K^j x_k + u_k)]^{\mathrm{T}} P[A_K x_k + B(K^j x_k + u_k)] \\ = & x_k^{\mathrm{T}} P^{j+1} x_k - x_k^{\mathrm{T}} A_K^{\mathrm{T}} P^{j+1} A_K x_k - (u_k + K^j x_k)^{\mathrm{T}} B^{\mathrm{T}} P^{j+1} B(u_k + K^j x_k) \\ & - 2 x_k^{\mathrm{T}} A_K^{\mathrm{T}} P^{j+1} B(K^j x_k + u_k) \\ = & x_k^{\mathrm{T}} P^{j+1} x_k - x_k^{\mathrm{T}} A_K^{\mathrm{T}} P^{j+1} A_K x_k - (u_k + K^j x_k)^{\mathrm{T}} B^{\mathrm{T}} P^{j+1} A_K x_k \\ & - (u_k + K^j x_k)^{\mathrm{T}} B^{\mathrm{T}} P^{j+1} [A_K x_k + B(K^j x_k + u_k)] \\ = & x_k^{\mathrm{T}} P^{j+1} x_k - x_k^{\mathrm{T}} A_K^{\mathrm{T}} P^{j+1} A_K x_k - (u_k + K^j x_k)^{\mathrm{T}} B^{\mathrm{T}} P^{j+1} A_K x_k \\ & - (u_k + K^j x_k)^{\mathrm{T}} B^{\mathrm{T}} P^{j+1} x_{k+1}. \end{aligned} \tag{4.14}$$

Due to (4.6), we also have

$$P = Q + K^{\mathrm{T}} R K + A_K^{\mathrm{T}} P A_K. \tag{4.15}$$

Substituting Eqs. (4.5) and (4.10) into Eq. (4.14) yields the off-policy Bellman equation

$$\begin{aligned} & x_k^{\mathrm{T}} P^{j+1} x_k - x_{k+1}^{\mathrm{T}} P^{j+1} x_{k+1} \\ = & x_k^{\mathrm{T}} Q x_k + x_k^{\mathrm{T}} (K^j)^{\mathrm{T}} R K^j x_k - (u_k + K^j x_k)^{\mathrm{T}} B^{\mathrm{T}} P^{j+1} A_K x_k \\ & - (u_k + K^j x_k)^{\mathrm{T}} B^{\mathrm{T}} P^{j+1} x_{k+1}. \end{aligned} \tag{4.16}$$

Using the property of Kronecker product, $x^{\mathrm{T}}Ay = (y^{\mathrm{T}} \otimes x^{\mathrm{T}})\mathrm{vec}(A)$, and considering the reconstructed dynamics (4.11), Eq. (4.16) can be rewritten as Eq. (4.13), which concludes the proof. □

Set $W_1^{j+1} = P^{j+1} \in \mathbb{R}^{n\times n}$, $W_2^{j+1} = B^{\mathrm{T}}P^{j+1}A \in \mathbb{R}^{m\times n}$, and $W_3^{j+1} = B^{\mathrm{T}}P^{j+1}B \in \mathbb{R}^{m\times m}$, which can be solved by least squares (LS) methods through (4.13). Thus P^{j+1} and K^{j+1} can be solved simultaneously without knowing the dynamics. There are $n^2 + m^2 + mn$ unknown parameters. So the window size for collecting data is selected as $s \geq n^2 + m^2 + mn$. Note that W_1^j and W_3^j are symmetric matrices with $n \times (n+1)/2$ and $m \times (m+1)/2$ independent elements, respectively. Hence only $n \times (n+1)/2$ and $m \times (m+1)/2$ elements need to be computed for W_1^j and W_3^j, respectively. Algorithm 4.3 describes the data-driven off-policy RL (Kiumarsi et al., 2017).

The convergence of Algorithm 4.2 is proved in Kiumarsi et al. (2017) by showing that the off-policy Bellman equation (4.12) is equivalent to the model-based Bellman equation (4.6). Thus, as j goes to ∞, P^{j+1} converges to the optimal kernel matrix $P^\star$, which is the solution of the ARE (4.9), and K^{j+1} converges to the optimal controller gain $K^\star$ given by (4.8). Step 1 in Algorithm 4.2 is implemented by the LS method

$$\zeta^{j+1} = ((\psi^j)^{\mathrm{T}}\psi^j)^{-1}(\psi^j)^{\mathrm{T}}\phi^j \tag{4.17}$$

with $\zeta^{j+1} = [\mathrm{vec}(W_1^{j+1})^{\mathrm{T}}\ \mathrm{vec}(W_2^{j+1})^{\mathrm{T}}\ \mathrm{vec}(W_3^{j+1})^{\mathrm{T}}\ ]^{\mathrm{T}}$ and the window size $s \geq n^2 + m^2 + mn$. PE conditions are required in (4.17) to guarantee ψ^j has full rank (Kiumarsi et al., 2017).

4.2.4 Data-driven off-policy RL with clock offsets

Now we consider data-driven off-policy RL under clock offsets. We observe from Algorithm 4.2 that clock offsets among different components influence the matrix formulation of ϕ and ψ. Under clock offsets, the state-input data actually utilized by the learning component at the time instant k are $x_k^l = x_{k+\delta_x(k)}$ and $u_k^l = u_{k+\delta_u(k)}$, where $\delta_x(k)$ is the clock offset of state signals, and $\delta_u(k)$ is the clock offset of control input signals. Then the matrix formulation under clock offsets is given by

$$\hat{\phi}^j = \begin{bmatrix} (x_k^l)^{\mathrm{T}}Qx_k^l + (x_k^l)^{\mathrm{T}}(K^j)^{\mathrm{T}}RK^jx_k^l \\ (x_{k+1}^l)^{\mathrm{T}}Qx_{k+1}^l + (x_{k+1}^l)^{\mathrm{T}}(K^j)^{\mathrm{T}}RK^jx_{k+1}^l \\ \vdots \\ (x_{k+s-1}^l)^{\mathrm{T}}Qx_{k+s-1}^l + (x_{k+s-1}^l)^{\mathrm{T}}(K^j)^{\mathrm{T}}RK^jx_{k+s-1}^l \end{bmatrix}, \tag{4.18}$$

Algorithm 4.2 Data-Driven Off-Policy RL.

Select a stabilizing control policy $u_k = \mu_b(x_k)$ for data collection. Set the iteration number $j = 0$. Select an initial controller gain K^0 and a proper window size s.

1. Solve for W_1^{j+1}, W_2^{j+1}, W_3^{j+1} such that, $\forall k \in \mathbb{N}$,

$$\psi^j[\text{vec}(W_1^{j+1})^{\mathrm{T}}\ \text{vec}(W_2^{j+1})^{\mathrm{T}}\ \text{vec}(W_3^{j+1})^{\mathrm{T}}]^{\mathrm{T}} = \phi^j,$$

with $\psi^j \in \mathbb{R}^{s\times(n^2+m^2+mn)}$ and $\phi^j \in \mathbb{R}^s$,

$$\phi^j = \begin{bmatrix} x_k^{\mathrm{T}} Q x_k + x_k^{\mathrm{T}}(K^j)^{\mathrm{T}} R K^j x_k \\ x_{k+1}^{\mathrm{T}} Q x_{k+1} + x_{k+1}^{\mathrm{T}}(K^j)^{\mathrm{T}} R K^j x_{k+1} \\ \vdots \\ x_{k+s-1}^{\mathrm{T}} Q x_{k+s-1} + x_{k+s-1}^{\mathrm{T}}(K^j)^{\mathrm{T}} R K^j x_{k+s-1} \end{bmatrix},$$

$$\psi^j = \begin{bmatrix} H_1^{\mathrm{xx}} & H_1^{\mathrm{xu}} & H_1^{\mathrm{uu}} \\ H_2^{\mathrm{xx}} & H_2^{\mathrm{xu}} & H_2^{\mathrm{uu}} \\ \vdots & \vdots & \vdots \\ H_s^{\mathrm{xx}} & H_s^{\mathrm{xu}} & H_s^{\mathrm{uu}} \end{bmatrix},$$

with, for $i = 1, 2, \ldots, s$,

$$\begin{aligned} H_i^{\mathrm{xx}} &= x_{k+i-1}^{\mathrm{T}} \otimes x_{k+i-1}^{\mathrm{T}} - x_{k+i}^{\mathrm{T}} \otimes x_{k+i}^{\mathrm{T}}, \\ H_i^{\mathrm{xu}} &= 2[x_{k+i-1}^{\mathrm{T}} \otimes (u_{k+i-1} + K^j x_{k+i-1})^{\mathrm{T}}], \\ H_i^{\mathrm{uu}} &= (u_{k+i-1} - K^j x_{k+i-1})^{\mathrm{T}} \otimes (u_{k+i-1} + K^j x_{k+i-1})^{\mathrm{T}}. \end{aligned}$$

2. Update policy by $K^{j+1} = (R + W_3^{j+1})^{-1} W_2^{j+1}$.
3. Stop if $\|K^{j+1} - K^j\| \le \xi$ with a positive threshold ξ. Otherwise, let $j = j+1$ and go to step 1.

$$\hat{\psi}^j = \begin{bmatrix} \hat{H}_1^{\mathrm{xx}} & \hat{H}_1^{\mathrm{xu}} & \hat{H}_1^{\mathrm{uu}} \\ \hat{H}_2^{\mathrm{xx}} & \hat{H}_2^{\mathrm{xu}} & \hat{H}_2^{\mathrm{uu}} \\ \vdots & \vdots & \vdots \\ \hat{H}_s^{\mathrm{xx}} & \hat{H}_s^{\mathrm{xu}} & \hat{H}_s^{\mathrm{uu}} \end{bmatrix}, \tag{4.19}$$

with, for $i = 1, 2, \ldots, s$,

$$\hat{H}_i^{\mathrm{xx}} = (x_{k+i-1}^l)^{\mathrm{T}} \otimes (x_{k+i-1}^l)^{\mathrm{T}} - (x_{k+i}^l)^{\mathrm{T}} \otimes (x_{k+i}^l)^{\mathrm{T}},$$

$$\hat{H}_i^{\text{xu}} = 2[(x_{k+i-1}^l)^{\text{T}} \otimes (u_{k+i-1}^l + K^j x_{k+i-1}^l)^{\text{T}}],$$
$$\hat{H}_i^{\text{uu}} = (u_{k+i-1}^l - K^j x_{k+i-1}^l)^{\text{T}} \otimes (u_{k+i-1}^l + K^j x_{k+i-1}^l)^{\text{T}}.$$

Accordingly, Algorithm 4.2 is replaced by Algorithm 4.3.

Algorithm 4.3 Data-Driven Off-Policy RL with Clock Offsets.

Select a stabilizing control policy $u_k = \mu_{\text{b}}(x_k)$ for data collection. Set the iteration number $j = 0$. Select an initial controller gain K^0 and a proper window size s.

1. Solve for $\hat{W}_1^{j+1}$, $\hat{W}_2^{j+1}$, $\hat{W}_3^{j+1}$ such that, $\forall k \in \mathbb{N}$,

$$\hat{\psi}^j[\text{vec}(\hat{W}_1^{j+1})^{\text{T}} \ \text{vec}(\hat{W}_2^{j+1})^{\text{T}} \ \text{vec}(\hat{W}_3^{j+1})^{\text{T}}]^{\text{T}} = \hat{\phi}^j,$$

where $\hat{\phi}^j \in \mathbb{R}^s$ and $\hat{\psi}^j \in \mathbb{R}^{s\times(n^2+m^2+mn)}$ are given by (4.18) and (4.19), respectively.

2. Update policy by

$$\hat{K}^{j+1} = (R + \hat{W}_3^{j+1})^{-1}\hat{W}_2^{j+1}. \tag{4.20}$$

3. Stop if $\|\hat{K}^{j+1} - \hat{K}^j\| \leq \xi$ with a positive threshold ξ. Otherwise, let $j = j+1$ and go to step 1.

Denote intermediate variables at the jth iteration as $\hat{W}_1^{j+1} = \hat{P}^{j+1}$, $\hat{W}_2^{j+1} = B^{\text{T}}\hat{P}^{j+1}A$, $\hat{W}_3^{j+1} = B^{\text{T}}\hat{P}^{j+1}B$. The LS estimation of ζ under clock offsets is given by

$$\hat{\zeta}^{j+1} = ((\hat{\psi}^j)^{\text{T}}\hat{\psi}^j)^{-1}(\hat{\psi}^j)^{\text{T}}\hat{\phi}^j. \tag{4.21}$$

4.2.5 Impact of clock offsets on data-driven off-policy RL

Now we study the impact of clock offsets on the learning behavior: with $\hat{\phi}^j$ and $\hat{\psi}^j$ whether Algorithm 4.3 still generates stabilizing control policies.

Lemma 4.2. *Assume that the PE condition is satisfied when collecting data for Algorithms 4.2 and 4.3. Then the control policies generated by Algorithm 4.3 are stabilizing policies given that the norm of the learning gap* $\|\epsilon\| = \|\zeta^{j+1} - \hat{\zeta}^{j+1}\|$ *is sufficiently small.*

Proof. Define the operator $\mathcal{H}$ as

$$\mathcal{H}(K, P, A, B, Q, R, x_k^l) = (x_k^l)^{\text{T}}[(A - BK)^{\text{T}}P(A - BK) + Q + K^{\text{T}}RK]x_k^l.$$

It follows from the model-based Bellman equation (4.6) that $\mathcal{H}(K, P, A, B, Q, R, x_k^l) = (x_k^l)^\mathrm{T} P x_k^l$. In the context of the jth iteration during the learning process, based on (4.10) from the model-based RL Algorithm 4.1, we have $\mathcal{H}(K^j, P^{j+1}, A, B, Q, R, x_k^l) = (x_k^l)^\mathrm{T} P^{j+1} x_k^l$. According to the Rayleigh–Ritz inequality for symmetric matrices, we have $\underline{\lambda}(P^{j+1} - \hat{P}^{j+1})\|x_k^l\|^2 \leq (x_k^l)^\mathrm{T} P^{j+1} x_k^l - (x_k^l)^\mathrm{T} \hat{P}^{j+1} x_k^l \leq \bar{\lambda}(P^{j+1} - \hat{P}^{j+1})\|x_k^l\|^2$. Then it follows that

$$|(x_k^l)^\mathrm{T} P^{j+1} x_k^l - (x_k^l)^\mathrm{T} \hat{P}^{j+1} x_k^l| \leq \varepsilon_1 \tag{4.22}$$

with $\varepsilon_1 = \max(|\underline{\lambda}(P^{j+1} - \hat{P}^{j+1})|\|x_k^l\|^2, \ |\bar{\lambda}(P^{j+1} - \hat{P}^{j+1})|\|x_k^l\|^2)$. Let $V^{j+1} = (A - BK^j)^\mathrm{T} P^{j+1}(A - BK^j) + (K^j)^\mathrm{T} R K^j$ and $\hat{V}^{j+1} = (A - B\hat{K}^j)^\mathrm{T} \hat{P}^{j+1}(A - B\hat{K}^j) + (\hat{K}^j)^\mathrm{T} R \hat{K}^j$. Following the same logic, we have

$$\begin{aligned}
&\underline{\lambda}(V^{j+1} - \hat{V}^{j+1})\|x_k^l\|^2 \\
&\leq \mathcal{H}(K^j, P^{j+1}, A, B, Q, R, x_k^l) - \mathcal{H}(\hat{K}^j, \hat{P}^{j+1}, A, B, Q, R, x_k^l) \\
&\leq \bar{\lambda}(V^{j+1} - \hat{V}^{j+1})\|x_k^l\|^2.
\end{aligned}$$

Then

$$|\mathcal{H}(K^j, P^{j+1}, A, B, Q, R, x_k^l) - \mathcal{H}(\hat{K}^j, \hat{P}^{j+1}, A, B, Q, R, x_k^l)| \leq \varepsilon_2 \tag{4.23}$$

with $\varepsilon_2 \leq \max(|\underline{\lambda}(V^{j+1} - \hat{V}^{j+1})|, |\bar{\lambda}(V^{j+1} - \hat{V}^{j+1})|)\|x_k^l\|^2$. According to (4.7) and (4.8), we know that the update control policy (4.20) in each iteration from Algorithm 4.3 is in fact the minimizer of $\mathcal{H}$ operator, i.e.,

$$\hat{K}^{j+1} = \arg\min_{\hat{K}} \mathcal{H}(\hat{K}, \hat{P}^{j+1}, A, B, Q, R, x_k^l).$$

Considering bounds ε_1 and ε_2 due to the learning gap ϵ from clock offsets,

$$\begin{aligned}
&\mathcal{H}(\hat{K}^{j+1}, \hat{P}^{j+1}, A, B, Q, R, x_k^l) \\
&< \mathcal{H}(\hat{K}^j, \hat{P}^{j+1}, A, B, Q, R, x_k^l) \\
&\leq \mathcal{H}(K^j, P^{j+1}, A, B, Q, R, x_k^l) + \varepsilon_2 \\
&= (x_k^l)^\mathrm{T} P^{j+1} x_k^l + \varepsilon_2 \\
&= (x_k^l)^\mathrm{T} \hat{P}^{j+1} x_k^l + \varepsilon_1 + \varepsilon_2.
\end{aligned}$$

Then we get

$$\begin{aligned}
&\mathcal{H}(\hat{K}^{j+1}, \hat{P}^{j+1}, A, B, Q, R, x_k^l) \\
&= (x_k^l)^\mathrm{T}[(A - B\hat{K}^{j+1})^\mathrm{T} \hat{P}^{j+1}(A - B\hat{K}^{j+1}) + Q
\end{aligned}$$

$$+ (\hat{K}^{j+1})^{\mathrm{T}} R \hat{K}^{j+1}] x_k^l < (x_k^l)^{\mathrm{T}} \hat{P}^{j+1} x_k^l + \varepsilon_1 + \varepsilon_2.$$

Since $Q \succeq 0$ and $R \succ 0$, for sufficiently small learning gap ϵ and thus sufficiently small ε_1 and ε_2, we have $(A - B\hat{K}^{j+1})^{\mathrm{T}} \hat{P}^{j+1} (A - B\hat{K}^{j+1}) < \hat{P}^{j+1}$, which implies that the largest absolute value of the eigenvalues of $A - B\hat{K}^{j+1}$ is smaller than one, i.e., the spectral radius $\rho(A - B\hat{K}^{j+1}) < 1$. Thus $\hat{K}^{j+1}$ is a stabilizing control policy. □

As a consequence, we have the following corollary.

Corollary 4.1. *Given the bound ε_1 described by (4.22) and the bound ε_2 described by (4.23) due to the learning gap ϵ from clock offsets, assume that $\varepsilon_1 + \varepsilon_2 < (x_k^l)^{\mathrm{T}} [Q + (\hat{K}^{j+1})^{\mathrm{T}} R \hat{K}^{j+1}] x_k^l$ at each iteration of the learning process. Then the control policies generated by Algorithm 4.3 with clock offsets are stabilizing policies.*

Proof. It is shown in the proof of Lemma 4.2 that

$$(x_k^l)^{\mathrm{T}} [(A - B\hat{K}^{j+1})^{\mathrm{T}} \hat{P}^{j+1} (A - B\hat{K}^{j+1}) + Q + (\hat{K}^{j+1})^{\mathrm{T}} R \hat{K}^{j+1}] x_k^l$$
$$< (x_k^l)^{\mathrm{T}} \hat{P}^{j+1} x_k^l + \varepsilon_1 + \varepsilon_2.$$

It follows from $\varepsilon_1 + \varepsilon_2 < (x_k^l)^{\mathrm{T}} [Q + (\hat{K}^{j+1})^{\mathrm{T}} R \hat{K}^{j+1}] x_k^l$ that $(A - B\hat{K}^{j+1})^{\mathrm{T}} \hat{P}^{j+1} (A - B\hat{K}^{j+1}) < \hat{P}^{j+1}$, which implies $\hat{K}^{j+1}$ is a stabilizing controller gain at each iteration. □

Based on Lemma 4.2, now we derive a direct relationship between clock offsets and the performance of off-policy RL.

Theorem 4.1. *Assume the PE condition is satisfied given the collection of data for Algorithms 4.2 and 4.3. Then the control policies generated by Algorithm 4.3 are stabilizing policies given that $\|A^{\delta_x(k)} - I_n\| \mathcal{X} + \| \Sigma_{i=0}^{\delta_x(k)-1} A^i B \| \mathcal{U}$ is sufficiently small for the case of $\delta_x(k) > 0$, or $\|A^{\delta_x(k)} - I_n\| \mathcal{X} + \| \Sigma_{i=0}^{-\delta_x(k)-1} A^{i+\delta_x(k)} B \|\ \mathcal{U}$ is sufficiently small for the case of $\delta_x(k) < 0$, where $\mathcal{X}$ is the upper bound of state norm, and $\mathcal{U}$ is the upper bound of input norm.*

Proof. It follows from $\psi^j \zeta^{j+1} = \phi^j$ and $\hat{\psi}^j \hat{\zeta}^{j+1} = \hat{\phi}^j$ at the jth iteration of Algorithms 4.2 and 4.3 that

$$\hat{\psi}^j \epsilon^{j+1} = \Delta \psi^j \zeta^{j+1} - \Delta \phi^j$$

with $\Delta \phi^j = \phi^j - \hat{\phi}^j$, $\Delta \psi^j = \psi^j - \hat{\psi}^j$, and $\epsilon^{j+1} = \zeta^{j+1} - \hat{\zeta}^{j+1}$; ϵ is the learning gap which is the difference between the real intermediate variable ζ from

Algorithm 4.2 and the corrupted one $\hat{\zeta}$ from Algorithm 4.3 under clock offsets. Given that $\hat{\psi}^j$ is invertible under the PE condition, if we could prove that $\Delta\psi^j\zeta^{j+1}$ and $\Delta\phi^j$ are sufficiently small under certain conditions and so $\|\epsilon^{j+1}\|$ is, then according to Lemma 4.2, Algorithm 4.3 could generate stabilizing control policies.

Consider the difference of state-input data generated with clock offsets and without clock offsets first. Manipulating dynamics (4.1) recursively, we get

$$x_k = A^k x_0 + \Sigma_{i=0}^{k-1} A^i B u_{k-1-i}. \tag{4.24}$$

Accordingly, state data with clock offset $\delta_x(k)$ are given by

$$x_k^l = x_{k+\delta_x(k)} = A^{k+\delta_x(k)} x_0 + \Sigma_{i=0}^{k+\delta_x(k)-1} A^i B u_{k+\delta_x(k)-1-i}. \tag{4.25}$$

For $\delta_x(k) > 0$, combining (4.24) and (4.25), we get

$$x_k^l - x_k = (A^{\delta_x(k)} - I_n)x_k + \Sigma_{i=0}^{\delta_x(k)-1} A^i B u_{k+\delta_x(k)-1-i}. \tag{4.26}$$

Note that the second term on the right-hand side of (4.26) contains a sum of finite terms and thus is bounded. Thus the norm of difference between state data with clock offsets and those without clock offsets are bounded by

$$\begin{aligned} \|x_k^l - x_k\| &\le \|A^{\delta_x(k)} - I_n\| \|x_k\| + \|\Sigma_{i=0}^{\delta_x(k)-1} A^i B u_{k+\delta_x(k)-1-i}\| \\ &\le \|A^{\delta_x(k)} - I_n\| \mathcal{X} + \|\Sigma_{i=0}^{\delta_x(k)-1} A^i B\| \mathcal{U}, \end{aligned} \tag{4.27}$$

where $\mathcal{X}$ is the upper bound of state norm, and $\mathcal{U}$ is the upper bound of input norm. Hence, if $\|A^{\delta_x(k)} - I_n\|\mathcal{X} + \|\Sigma_{i=0}^{\delta_x(k)-1} A^i B\|\mathcal{U}$ is sufficiently small, then $\|x_k^l - x_k\|$ is sufficiently small. Likewise, for $\delta_x(k) < 0$, it follows from (4.24) and (4.25) that

$$\begin{aligned} \|x_k^l - x_k\| &= \|(A^{\delta_x(k)} - I_n)x_k - \Sigma_{i=0}^{-\delta_x(k)-1} A^{i+\delta_x(k)} B u_{k-1-i}\| \\ &\le \|A^{\delta_x(k)} - I_n\| \|x_k\| + \|\Sigma_{i=0}^{-\delta_x(k)-1} A^{i+\delta_x(k)} B u_{k-1-i}\| \\ &\le \|A^{\delta_x(k)} - I_n\| \mathcal{X} + \|\Sigma_{i=0}^{-\delta_x(k)-1} A^{i+\delta_x(k)} B\| \mathcal{U}. \end{aligned} \tag{4.28}$$

Combining (4.27) and (4.28), we get

$$\begin{aligned} \|x_k^l - x_k\| &\le \|A^{\delta x(k)} - I_n\| \mathcal{X} \\ &+ \max(\|\Sigma_{i=0}^{\delta_x(k)-1} A^i B\|, \|\Sigma_{i=0}^{-\delta_x(k)-1} A^{i+\delta_x(k)} B\|) \mathcal{U}. \end{aligned} \tag{4.29}$$

For the stabilizing control policy $u_k = \mu_b(x_k)$ during data collection phase, let the state corresponding to $u_k^l = u_{k+\delta_u(k)}$ be $\bar{x}_k$. Denote $\bar{x}_k = x_{k+\delta_{\bar{u}}(k)}$, where $\delta_{\bar{u}}(k)$ is a function of $\delta_u(k)$. According to the Lipschitz continuity and (4.29), we have

$$\|u_k^l - u_k\| = \|\mu_b(\bar{x}_k) - \mu_b(x_k)\| \leq L\|x_{k+\delta_{\bar{u}}(k)} - x_k\|,$$

where L is a Lipschitz constant for the function μ_b. Therefore, if $\|A^{\delta x(k)} - I_n\|\mathcal{X} + \max(\|\Sigma_{i=0}^{\delta_x(k)-1} A^i B\|, \|\Sigma_{i=0}^{-\delta_x(k)-1} A^{i+\delta_x(k)} B\|)\mathcal{U}$ is sufficiently small, then the norm of difference of input data between with clock offsets and without clock offsets $\|u_k^l - u_k\|$ is sufficiently small.

For the ith row and 1st column term of $\Delta\psi^j\zeta^{j+1}$, where $1 \leq i \leq s$, we have

$$\begin{aligned}
&(H_i^{xx} - \hat{H}_i^{xx})\text{vec}(W_1^{j+1}) \\
&= x_{k+i-1}^{\text{T}} W_1^{j+1} x_{k+i-1} - x_{k+i}^{\text{T}} W_1^{j+1} x_{k+i} \\
&\quad - [(x_{k+i-1}^l)^{\text{T}} W_1^{j+1} x_{k+i-1}^l - (x_{k+i}^l)^{\text{T}} W_1^{j+1} x_{k+i}^l] \\
&= (x_{k+i-1} - x_{k+i-1}^l)^{\text{T}} W_1^{j+1} \left(x_{k+i-1} + (x_{k+i-1}^l)\right) \\
&\quad + (x_{k+i} - x_{k+i}^l)^{\text{T}} W_1^{j+1} \left(x_{k+i} + (x_{k+i}^l)\right).
\end{aligned}$$

It follows that

$$\begin{aligned}
&\|(H_i^{xx} - \hat{H}_i^{xx})\text{vec}(W_1^{j+1})\| \\
&\leq 2(\|x_{k+i-1} - x_{k+i-1}^l)^{\text{T}}\| + \|(x_{k+i} - x_{k+i}^l)^{\text{T}}\|)\|W_1^{j+1}\|\mathcal{X}.
\end{aligned}$$

If $\|x_k^l - x_k\|$ is sufficiently small, then $(H_i^{xx} - \hat{H}_i^{xx})\text{vec}(W_1^{j+1})$ is sufficiently small.

For the ith row and 2nd column term of $\Delta\psi^j\zeta^{j+1}$, where $1 \leq i \leq s$, we have

$$\begin{aligned}
&\|(H_i^{xu} - \hat{H}_i^{xu})\text{vec}(W_2^{j+1})\| \\
&= \|2[u_{k+i-1}^{\text{T}} W_2^{j+1} x_{k+i-1} - (u_{k+i-1}^l)^{\text{T}} W_2^{j+1} x_{k+i-1}^l] \\
&\quad + 2[x_{k+i-1}^{\text{T}} (K^j)^{\text{T}} W_2^{j+1} x_{k+i-1} \\
&\quad - (x_{k+i-1}^l)^{\text{T}} (K^j)^{\text{T}} W_2^{j+1} x_{k+i-1}^l]\| \\
&= \|2[u_{k+i-1}^{\text{T}} W_2^{j+1} (x_{k+i-1} - x_{k+i-1}^l) \\
&\quad + (x_{k+i-1}^l)^{\text{T}} (W_2^{j+1})^{\text{T}} (u_{k+i-1} - u_{k+i-1}^l)]
\end{aligned}$$

$$
\begin{aligned}
&+2[x_{k+i-1}^{\mathrm{T}}(K^j)^{\mathrm{T}}W_2^{j+1}(x_{k+i-1}-x_{k+i-1}^l)+\\
&+(x_{k+i-1}^l)^{\mathrm{T}}(W_2^{j+1})^{\mathrm{T}}K^j(x_{k+i-1}-x_{k+i-1}^l)]\|\\
\leq{}& 2\mathcal{U}\|W_2^{j+1}\|\cdot\|x_{k+i-1}-x_{k+i-1}^l\|\\
&+2\mathcal{X}\|(W_2^{j+1})^{\mathrm{T}}\|\cdot\|u_{k+i-1}-u_{k+i-1}^l\|\\
&+2\mathcal{X}\|(K^j)^{\mathrm{T}}W_2^{j+1}\|\cdot\|x_{k+i-1}-x_{k+i-1}^l\|\\
&+2\mathcal{X}\|(W_2^{j+1})^{\mathrm{T}}K^j\|\cdot\|x_{k+i-1}-x_{k+i-1}^l\|.
\end{aligned}
$$

If $\|x_k^l-x_k\|$ is sufficiently small, then $(H_i^{\mathrm{xu}}-\hat{H}_i^{\mathrm{xu}})\mathrm{vec}(W_2^{j+1})$ is sufficiently small.

For the ith row and 3rd column term of $\Delta\psi^j\zeta^{j+1}$, where $1\leq i\leq s$, we have

$$
\begin{aligned}
&\|(H_i^{\mathrm{uu}}-\hat{H}_i^{\mathrm{uu}})\mathrm{vec}(W_3^{j+1})\|\\
={}&\|[u_{k+i-1}^{\mathrm{T}}W_3^{j+1}u_{k+i-1}-(u_{k+i-1}^l)^{\mathrm{T}}W_3^{j+1}u_{k+i-1}^l]\\
&-[x_{k+i-1}^{\mathrm{T}}(K^j)^{\mathrm{T}}W_3^{j+1}K^jx_{k+i-1}\\
&-(x_{k+i-1}^l)^{\mathrm{T}}(K^j)^{\mathrm{T}}W_3^{j+1}K^jx_{k+i-1}^l]\\
&-[u_{k+i-1}^{\mathrm{T}}W_3^{j+1}K^jx_{k+i-1}-(u_{k+i-1}^l)^{\mathrm{T}}W_3^{j+1}K^jx_{k+i-1}^l]\\
&+[x_{k+i-1}^{\mathrm{T}}(K^j)^{\mathrm{T}}W_3^{j+1}u_{k+i-1}\\
&-(x_{k+i-1}^l)^{\mathrm{T}}(K^j)^{\mathrm{T}}W_3^{j+1}u_{k+i-1}^l]\|\\
={}&\|(u_{k+i-1}+u_{k+i-1}^l)^{\mathrm{T}}W_3^{j+1}(u_{k+i-1}-u_{k+i-1}^l)\\
&+(x_{k+i-1}+x_{k+i-1}^l)^{\mathrm{T}}(K^j)^{\mathrm{T}}W_3^{j+1}K^j(x_{k+i-1}-x_{k+i-1}^l)\\
&+u_{k+i-1}^{\mathrm{T}}W_3^{j+1}K^j(x_{k+i-1}-x_{k+i-1}^l)\\
&+(x_{k+i-1}^l)^{\mathrm{T}}(K^j)^{\mathrm{T}}W_3^{j+1}(u_{k+i-1}-u_{k+i-1}^l)\\
&+x_{k+i-1}^{\mathrm{T}}(K^j)^{\mathrm{T}}W_3^{j+1}(u_{k+i-1}-u_{k+i-1}^l)\\
&+(u_{k+i-1}^l)^{\mathrm{T}}W_3^{j+1}K^j(x_{k+i-1}-x_{k+i-1}^l)\|\\
\leq{}& 2\mathcal{U}W_3^{j+1}\|\cdot\|(u_{k+i-1}-u_{k+i-1}^l)\|\\
&+2\mathcal{X}\|(K^j)^{\mathrm{T}}W_3^{j+1}K^j\|\|(x_{k+i-1}-x_{k+i-1}^l)\|\\
&+2\mathcal{U}\|W_3^{j+1}K^j\|\cdot\|(x_{k+i-1}-x_{k+i-1}^l)\|\\
&+2\mathcal{X}\|(K^j)^{\mathrm{T}}W_3^{j+1}\|\cdot\|(u_{k+i-1}-u_{k+i-1}^l)\|.
\end{aligned}
$$

If $\|x_k^l-x_k\|$ is sufficiently small, then $(H_i^{\mathrm{uu}}-\hat{H}_i^{\mathrm{uu}})\mathrm{vec}(W_3^{j+1})$ is sufficiently small.

The ith term of $\Delta\phi^j$, where $1 \le i \le s$, is bounded by

$$\begin{aligned}
\|\Delta\phi_i^j\| &= \|x_{k+i-1}^{\mathrm{T}} Q x_{k+i-1} + x_{k+i-1}^{\mathrm{T}} (K^j)^{\mathrm{T}} R K^j x_{k+i-1} \| \\
&\quad - (x_{k+i-1}^l)^{\mathrm{T}} Q x_{k+i-1}^l - (x_{k+i-1}^l)^{\mathrm{T}} (\hat{K}^j)^{\mathrm{T}} R \hat{K}^j x_{k+i-1}^l \\
&= \|(x_{k+i-1} + x_{k+i-1}^l)^{\mathrm{T}} \left(Q + (K^j)^{\mathrm{T}} R K^j\right) (x_{k+i-1} - x_{k+i-1}^l) \\
&\quad + (x_{k+i-1}^l)^{\mathrm{T}} \left((K^j)^{\mathrm{T}} R K^j - (\hat{K}^j)^{\mathrm{T}} R \hat{K}^j\right) x_{k+i-1}^l \| \\
&\le 2\mathcal{X} \|Q + (K^j)^{\mathrm{T}} R K^j\| \, \|x_{k+i-1} - x_{k+i-1}^l\| \\
&\quad + \bar{\lambda}(Q + (K^j)^{\mathrm{T}} R K^j) \|x_{k+i-1} - x_{k+i-1}^l\|^2.
\end{aligned}$$

Thus, if $\|x_k^l - x_k\|$ is sufficiently small, then each term in $\Delta\phi^j$ is sufficiently small. This completes the proof. □

Remark 4.1. For the sufficient condition of generating stabilizing control policies by Algorithm 4.3, the first term $\|A^{\delta_x(k)} - I_n\|\mathcal{X}$ implies that the impact of clock offsets depends on the deviation of dynamics change $A^{\delta_x(k)}$ from the identity matrix I_n. The second term $\|\Sigma_{i=0}^{\delta_x(k)-1} A^i B\|\mathcal{U}$ implies that the impact of clock offsets depends on the control input during the duration of asynchronization. If the closed-loop dynamics does not change too fast and clock offsets are small enough, then the off-policy RL with clock offsets could still generate stabilizing control policies. The learning gap ϵ reflects the combined effects of system dynamics and clock offsets.

Now we analyze the performance loss incurred due to clock offsets. The optimal value at the time instant k is given by $V^\star(x_k) = x_k^{\mathrm{T}} P^\star x_k$ without clock offsets and $\hat{V}(x_k^l) = (x_k^l)^{\mathrm{T}} \hat{P} x_k^l$ with clock offsets.

Theorem 4.2. *Assume that a linear state feedback behavior policy K_b is applied to the system, and the matrix $(A - BK_b)$ is invertible. With the clock offset for state signals $\delta_x(k)$ and the clock offset for control input signals $\delta_u(k)$, the performance loss is bounded as $|V^\star(x_k) - \hat{V}(x_k^l)| \le \tilde{\epsilon}\|x_k\|^2$, where $\tilde{\epsilon} = \max(|\underline{\lambda}(P^\star - \hat{P})|, |\bar{\lambda}(P^\star - \hat{P})|) + \max(|\underline{\lambda}(\mathcal{P})|, |\bar{\lambda}(\mathcal{P})|)$ with $\mathcal{P} = \hat{P} - ((A - BK_b)^{\delta_x(k)})^{\mathrm{T}} \hat{P} (A - BK_b)^{\delta_x(k)}$.*

Proof. We can write

$$\begin{aligned}
&|V^\star(x_k) - \hat{V}(x_k^l)| \\
&= |x_k^{\mathrm{T}} P^\star x_k - (x_k^l)^{\mathrm{T}} \hat{P} x_k^l| \\
&= |x_k^{\mathrm{T}} P^\star x_k - x_k^{\mathrm{T}} \hat{P} x_k + x_k^{\mathrm{T}} \hat{P} x_k - (x_k^l)^{\mathrm{T}} \hat{P} x_k^l| \\
&\le |x_k^{\mathrm{T}} P^\star x_k - x_k^{\mathrm{T}} \hat{P} x_k| + |x_k^{\mathrm{T}} \hat{P} x_k - (x_k^l)^{\mathrm{T}} \hat{P} x_k^l|. \qquad (4.30)
\end{aligned}$$

Given the behavior policy K_{b}, we have

$$x_k^l = (A - BK_{\mathrm{b}})^{\delta_x(k)} x_k. \tag{4.31}$$

Denote $\mathcal{P} := \hat{P} - \left((A - BK_{\mathrm{b}})^{\delta_x(k)}\right)^{\mathrm{T}} \hat{P}(A - BK_{\mathrm{b}})^{\delta_x(k)}$. Substituting (4.31) into (4.30), according to the Rayleigh–Ritz inequality for symmetric matrices, we get

$$\begin{aligned}
&|V^\star(x_k) - \hat{V}(x_k^l)| \\
&\leq |x_k^{\mathrm{T}} P^\star x_k - x_k^{\mathrm{T}} \hat{P} x_k| + |x_k^{\mathrm{T}} \hat{P} x_k - (x_k^l)^{\mathrm{T}} \hat{P} x_k^l| \\
&= |x_k^{\mathrm{T}} P^\star x_k - x_k^{\mathrm{T}} \hat{P} x_k| + |x_k^{\mathrm{T}} \hat{P} x_k - x_k^{\mathrm{T}} \left((A - BK_{\mathrm{b}})^{\delta_x(k)}\right)^{\mathrm{T}} \hat{P}(A - BK_{\mathrm{b}})^{\delta_x(k)} x_k| \\
&\leq \max(|\underline{\lambda}(P^\star - \hat{P})|, \ |\bar{\lambda}(P^\star - \hat{P})|)\|x_k\|^2 + \max(|\underline{\lambda}(\mathcal{P})|, |\bar{\lambda}(\mathcal{P})|)\|x_k\|^2.
\end{aligned}$$

This completes the proof. □

4.2.6 Simulation results

To illustrate the proposed framework, we apply the data-driven off-policy RL Algorithms 4.2 and 4.3 to a third-order F-16 autopilot aircraft plant described by, for all $k \in \mathbb{N}$,

$$x_{k+1} = \begin{bmatrix} 0.9065 & 0.0816 & -0.0005 \\ 0.0743 & 09012 & -0.0007 \\ 0 & 0 & 0.1327 \end{bmatrix} x_k + \begin{bmatrix} -0.0027 \\ -0.0068 \\ 1 \end{bmatrix} u_k,$$

where the states are $x_k = [\alpha \ q \ \delta_e]^{\mathrm{T}}$ with the angle of attack α, the pitch rate q, and the elevator deflection angle δ_e. The control input u is the elevator actuator voltage. Select $Q = I_n$ and $R = I_m$. Solving the ARE (4.9), we get

$$P^\star = \begin{bmatrix} 15.1268 & 12.0315 & -0.0082 \\ 12.0315 & 12.2282 & -0.0080 \\ -0.0082 & -0.0080 & 1.0088 \end{bmatrix}$$

with the optimal state feedback control policy $K^\star = \begin{bmatrix} -0.0643 & 0.0699 & 0.0667 \end{bmatrix}$. Let the initial state be $x_0 = [10 \ \ -10 \ \ -3]$, and let the behavior policy be $K_{\mathrm{b}} = [0 \ 0.12 \ 1]$. To guarantee the richness of data, probing noise $e_k = \sin^2(0.5k) + \sin(k) + \cos(k)$ is added to the system for 50 time steps for data collection. The state trajectory with the behavior policy K_{b} during the data collection phase is shown in Fig. 4.2.

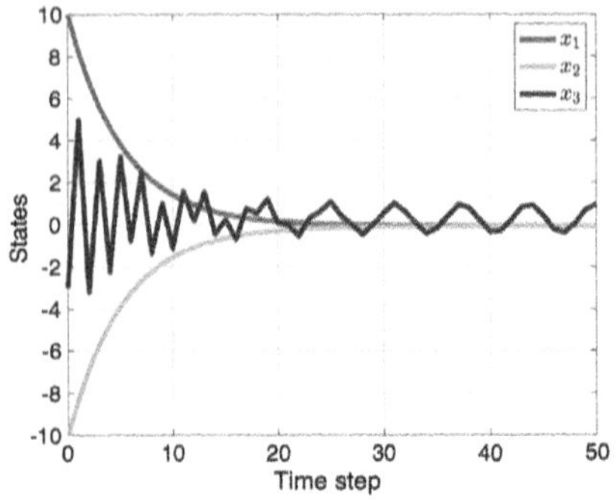

Figure 4.2 State trajectory with the behavior policy for data collection.

Algorithm 4.2 is implemented to learn the optimal control policy with the initial controller gain selected as $K^0 = [0\ 0.12\ 1]$. The window size for data collection should satisfy $s \geq n^2 + m^2 + mn = 13$ in this case, and we select $s = 16$ in the simulation. The norm of the difference between the iterative controller gains and the optimal feedback gain $K^\star$ and the norm of the difference between the iterative kernel matrices and the optimal kernel matrix $P^\star$ are shown in Fig. 4.3. Observe from the figure that K^j and P^j converge to the optimal values in less than 5 iterations with the converged values $P = \begin{bmatrix} 15.1268 & 12.0315 & -0.00822 \\ 12.0315 & 15.2282 & -0.0080 \\ -0.0082 & -0.0080 & 1.0088 \end{bmatrix}$ and $K = \begin{bmatrix} -0.0643 & -0.0699 & 0.0667 \end{bmatrix}$. Thus the efficacy of the data-driven off-policy RL Algorithm 4.2 is verified.

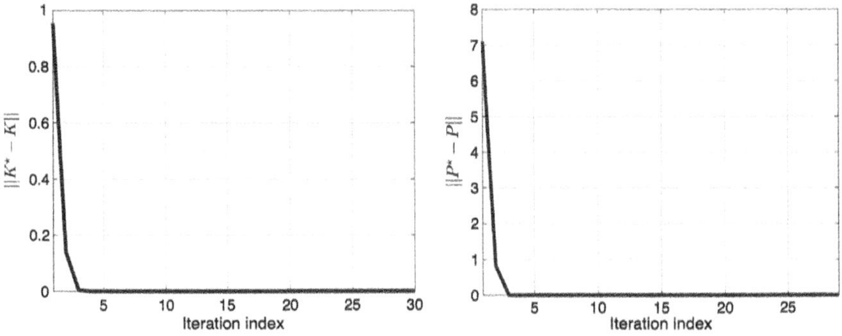

Figure 4.3 Learning process of the data-driven off-policy RL Algorithm 4.2.

Next, we study the influence of clock offsets on the data-driven off-policy RL by implementing Algorithm 4.3. Consider clock offsets between the controller and the learning component. The control input signals received by the learning component are actually $u_{k+\delta_u(k)}$ at the time instant k with $\delta_u = \{-1, -2, -3, -4, -5, -6, -7\}$. The data used for Algorithm 4.3 begin from the time step $k = 8$ with window size $s = 16$. The learning results under clock offsets are shown in Fig. 4.4. The state trajectories with learned policies are shown in Fig. 4.5. Observe from Fig. 4.4 that all the controller gains K^j and the kernel matrices P^j converge in the seven clock offset cases. However, note that in Fig. 4.5 the learned controller gain with clock offset $\delta_u = 1$ makes the system unstable and states explode, whereas the learned controller gain with bigger clock offset $\delta_u = 7$ stabilizes the system. As shown in Theorem 4.1, larger values of clock offsets of network nodes in CPS do not necessarily lead to negative influence on learning algorithms, i.e., generating non-stabilizing control policies. Both system dynamics and clock offsets determine the influence.

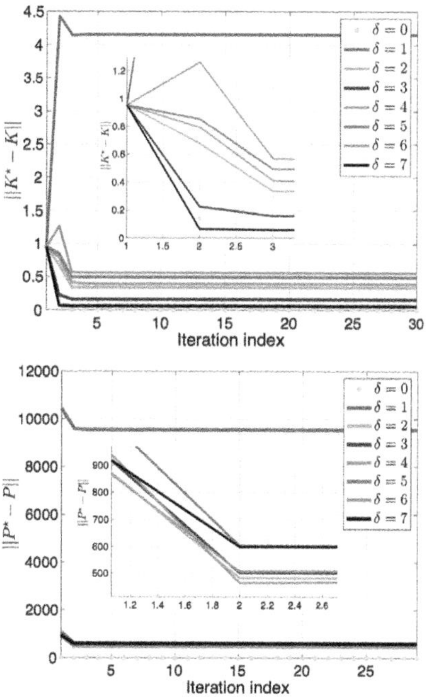

Figure 4.4 Learning process of Algorithm 4.3 under various clock offsets.

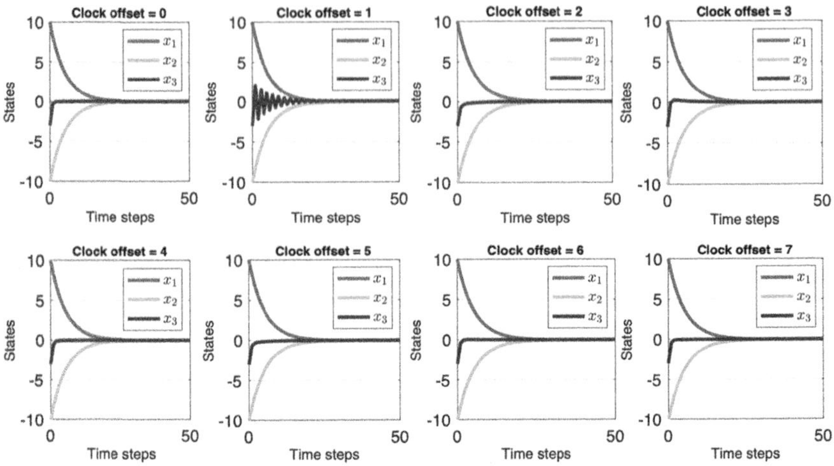

Figure 4.5 State trajectories with learned policies under clock offsets of control input signals with window size $s = 16$.

Table 4.1 Learning gaps with clock offsets of control input signals.

δ_u	1	2	3	4	5	6	7
ϵ	8261	417	435	441	522	404	520

From Section 4.2.5 we know that the learning gap ϵ from (4.21) directly determines the influence of clock offsets on learning algorithms. Learning gaps for the seven clock offset cases are presented in Table 4.1. Note that the learning gap for clock offset $\delta_u = 1$ is much larger than other cases, which explains the stabilizing differences of learned controller policies among these cases. According to (4.21), which part of data used for the data-driven off-policy RL Algorithm 4.3 has an impact on the learning gap ϵ under clock offsets. For example, if the data used for Algorithm 4.3 begin from $k = 8$ with window size $s = 40$, then for the clock offset $\delta_u = 1$, the learned controller gain is a stabilizing policy as shown in Fig. 4.6, which can be explained by Theorem 4.1 that both system dynamics and the magnitude of clock offsets together influence the learning behavior.

4.3. Clock offsets during autonomous decision-making for continuous-time systems

In this section, we shift our focus on the impact of clock offsets on the data-driven control of continuous-time systems.

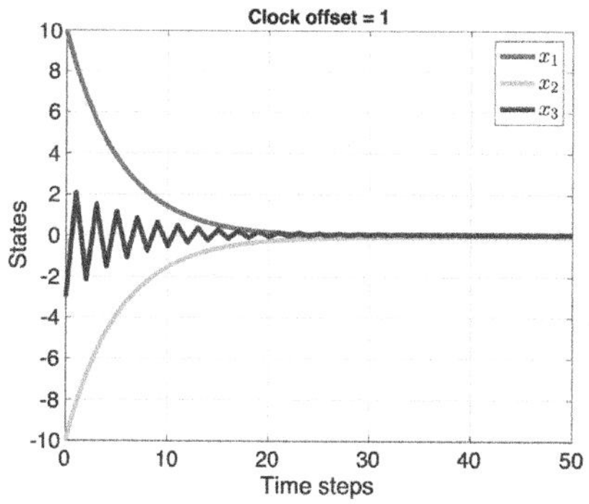

Figure 4.6 State trajectory with the learned policy under $\delta_u = 1$ with $s = 40$.

4.3.1 Characterization of clock offsets

Consider, for all $t \geq t_0$, a continuous-time nonlinear system of the form

$$\begin{aligned} \dot{x}(t) &= f(x(t)) + g(x(t))u(t) + h(x(t))a(t), \\ x(t_0) &= x_0, \end{aligned} \tag{4.32}$$

where $x(t) \in \mathbb{R}^n$ is the system state, $u(t) \in \mathbb{R}^m$ is the control input, $a(t) \in \mathbb{R}^p$ is an adversarial input, $f : \mathbb{R}^n \to \mathbb{R}^n$ is the system drift dynamics function, and $g : \mathbb{R}^n \to \mathbb{R}^{n\times m}$ and $h : \mathbb{R}^n \to \mathbb{R}^{n\times p}$ are the system input matrix functions. The functions f, g, and h are considered to be locally Lipschitz, though unknown. In addition, we assume that $f(0) = 0$, so that the origin is an equilibrium point of the uncontrolled system.

In complex CPS, which are systems of high heterogeneity, the presence of multiple physical and virtual components is almost certain. As a result of this distributed nature of CPS, it is extraordinarily difficult to design a unique, centralized clock, with which every system component will be perfectly synchronized (Shrivastava et al., 2016). In fact, it is more customary for every component, or a set of components, to have their own clock and a slightly different sense of time. Consequently, due to noise, quantizations, and other uncertainties common in practice (Okano et al., 2017), mismatches can occur between two or more component clocks.

In the mathematical expression of dynamics (4.32) of a CPS, we can observe several different kinds of CPS components: the state sensors, which measure the state $x(t)$ for $t \geq t_0$; the actuators, which implement the control policy $u(t)$ for $t \geq t_0$; and any potential exogenous input sensors, which measure $a(t)$ for $t \geq t_0$. To capture the general case regarding these components, let us consider that the CPS is equipped with n state sensors,

m actuators, and p exogenous input sensors. Then each of the sensors $i \in \{1, \ldots, n\} := \mathcal{N}_x$ provides us with measurements of the ith element of the state $x(t)$, which we denote as $\bar{x}_i(t)$, $t \geq t_0$. In addition, each actuator $j \in \{1, \ldots, m\} := \mathcal{N}_u$ provides us with values of the jth element of the control input $u(t)$, which we denote as $\bar{u}_j(t)$, $t \geq t_0$. Finally, each exogenous input sensor $l \in \{1, \ldots, p\} := \mathcal{N}_a$ transmits measured values of the exogenous input $a(t)$, which we denote as $\bar{a}_l(t)$, $t \geq t_0$.

Let us now define a "true/reference clock" $c : [t_0, \infty) \rightarrow [t_0, \infty)$, as the clock that is in agreement with the "real" time, that is, $c(t) = t$ for $t \geq t_0$. In case that the sensors and the actuators perceive real time accurately and are perfectly synchronized with one another as well as with the true clock, we will have

$$\begin{aligned}
\bar{x}_i(t) &= x_i(c(t)) = x_i(t), \ t \in [t_0, \infty), \ i \in \mathcal{N}_x, \\
\bar{u}_j(t) &= u_j(c(t)) = u_j(t), \ t \in [t_0, \infty), \ j \in \mathcal{N}_u, \\
\bar{a}_l(t) &= a_l(c(t)) = a_l(t), \ t \in [t_0, \infty), \ l \in \mathcal{N}_a.
\end{aligned}$$

Nevertheless, the clocks of the components of a CPS are rarely synchronized. As a result, it is more accurate to state that for all $t \in [t_0, \infty)$,

$$\begin{aligned}
\bar{x}_i(t) &= x_i(c_i^x(t)) = x_i(t + \delta_i^x(t)), \ i \in \mathcal{N}_x, \\
\bar{u}_j(t) &= u_j(c_j^u(t)) = u_j(t + \delta_j^u(t)), \ j \in \mathcal{N}_u, \\
\bar{a}_l(t) &= a_l(c_l^a(t)) = a_l(t + \delta_l^a(t)), \ l \in \mathcal{N}_a.
\end{aligned} \tag{4.33}$$

In (4.34), $c_i^x(t), c_j^u(t), c_l^a(t) \in [t_0, \infty)$ are the (strictly increasing) clock functions of each state sensor $i \in \mathcal{N}_x$, actuator $j \in \mathcal{N}_u$, and exogenous/adversarial input sensor $l \in \mathrm{N}_a$, respectively. Accordingly, the functions δ_i^x, δ_j^u, and δ_l^a are the *clock offsets* of each of these components from the reference clock. They are defined, for all $t \in [t_0, \infty)$, as

$$\begin{aligned}
\delta_i^x(t) &= c_i^x(t) - t, \ i \in \mathcal{N}_x, \\
\delta_j^u(t) &= c_j^u(t) - t, \ j \in \mathcal{N}_u, \\
\delta_l^a(t) &= c_l^a(t) - t, \ l \in \mathcal{N}_a.
\end{aligned} \tag{4.34}$$

When offsets (4.34) are nonzero, it is possible for a learning-based algorithm depending on the data (4.33) to become non-convergent or yield wrong results. Therefore it is of interest to investigate whether learning retains any kind of robustness toward clock offsets – at least, in the epsilon–delta sense. Here we will be specifically concerned about learning-based algorithms used to solve differential games in a model-free manner.

4.3.2 Two-player differential game

A zero-sum differential game can be defined over dynamics (4.32) where the competing players are: a) the CPS operator, aiming to optimally regulate the system; and b) a potential adversary or exogenous input, whose goal is assumed to be the disruption of the system performance. The utility functional of such a game can be defined as

$$J(x_0;\, u,\, a)=\int_{t_0}^{\infty}(Q(x(\tau))+u(\tau)^{\mathrm{T}}Ru(\tau)-\gamma^2 a(\tau)^{\mathrm{T}}a(\tau))\mathrm{d}\tau,$$

where $Q:\mathbb{R}^n\to\mathbb{R}$ is a positive definite function, $R\succ 0$, and $\gamma>0$ is an attenuation factor (Başar and Bernhard, 2008). The corresponding game is given by

$$V^\star(x)=\min_u\max_a J(x;\ u,\ a), \tag{4.35}$$

where $V^\star$ is the optimal value function. We consider that the game (4.35) has a unique solution, corresponding to a saddle point/Nash equilibrium. Hence

$$V^\star(x)=\min_u\max_a J(x;\ u,\ a)=\max_a\min_u J(x;\ u,\ a)$$

with the Nash/saddle-point/optimal policies given by

$$\{u^\star,\ a^\star\}=\arg\min_u\max_a J(\cdot;\ u,\ a).$$

In general, for a saddle-point policy tuple to exist, it is necessary that $\gamma>\gamma^\star$, where $\gamma^\star$ is the minimum attainable attenuation factor (Vamvoudakis and Lewis, 2012).

Now, note that a continuously differentiable value function V of a tuple $\{\mu_u,\ \mu_a\}$, where $\mu_u:\mathbb{R}^n\to\mathbb{R}^m$ and $\mu_a:\mathbb{R}^n\to\mathbb{R}^p$, must satisfy the Lyapunov-like nonlinear equation

$$\begin{aligned}\nabla V^{\mathrm{T}}(x)(f(x)+g(x)\mu_u(x)+h(x)\mu_a(x))\\ +Q(x)+\mu_u(x)^{\mathrm{T}}R\mu_u(x)-\gamma^2\mu_a(x)^{\mathrm{T}}\mu_a(x)=0.\end{aligned} \tag{4.36}$$

Additionally, given that $V^\star$ is continuously differentiable, following standard procedures (Lewis et al., 2012), the saddle-point $\{u^\star,\ a^\star\}$ of the game (4.35) can be shown to satisfy, for all $x\in\mathbb{R}^n$,

$$u^\star(x)=-\frac{1}{2}R^{-1}g^{\mathrm{T}}(x)\nabla V^\star(x),$$

$$a^\star(x) = \frac{1}{2\gamma^2} h^{\mathrm{T}}(x) \nabla V^\star(x).$$

Hence, plugging $u^\star$, $a^\star$, and $V^\star$ in the place of μ_u, μ_a, and V in (4.36), we obtain the Hamilton–Jacobi–Isaacs (HJI) equation

$$\begin{aligned} &\nabla V^{\star\mathrm{T}}(x) f(x) - \frac{1}{4} \nabla V^{\star\mathrm{T}}(x) g(x) R^{-1} g^{\mathrm{T}}(x) \nabla V^\star(x) \\ &+ \frac{1}{4\gamma^2} \nabla V^{\star\mathrm{T}}(x) h(x) h^{\mathrm{T}}(x) \nabla V^\star(x) + Q(x) = 0, \quad V^\star(0) = 0. \end{aligned} \tag{4.37}$$

To find the optimal policy $u^\star$, we need to solve the HJI equation (4.37), which is too difficult to be tackled analytically. Still, Algorithm 4.4, which describes the Policy Iteration (PI) procedure, can be used to calculate $V^\star$ through successive approximations over a given compact set $\Omega \subset \mathbb{R}^n$ (Wu and Luo, 2012).

Algorithm 4.4 Policy Iteration.

Let $i = 0$, $\Omega \subset \mathbb{R}^n$, $\epsilon > 0$. Start with a tuple of policies $\{u_0, a_0\}$ that is stabilizing in Ω.

repeat

Solve for V_i, $\forall x \in \Omega$, in

$$\begin{aligned} &\nabla V_i^{\mathrm{T}}(x)(f(x) + g(x) u_i(x) + h(x) a_i(x)) \\ &\qquad + Q(x) + u_i(x)^{\mathrm{T}} R u_i(x) - \gamma^2 a_i(x)^{\mathrm{T}} a_i(x) = 0, \end{aligned} \tag{4.38}$$

Let the new policies be given by

$$\begin{aligned} u_{i+1}(x) &= -\frac{1}{2} R^{-1} g^{\mathrm{T}}(x) \nabla V_i(x), \\ a_{i+1}(x) &= \frac{1}{2\gamma^2} h^{\mathrm{T}}(x) \nabla V_i(x). \end{aligned} \tag{4.39}$$

Set $i = i + 1$.

until $i \geq 2$ & $\sup_{x \in \Omega} |V_{i-1}(x) - V_{i-2}(x)| < \epsilon$.

4.3.3 Learning-based PI

Although Algorithm 4.4 can be used to effectively solve the HJI equation, it has the drawback of being model-based. In particular, we can see that the values of the functions f, g, and h over Ω are needed to solve the corresponding Lyapunov-like equation. To tackle this issue, Jiang and Jiang (2014); Modares et al. (2015) proposed a learning-based PI algorithm, which can approximate the optimal value function $V^\star$ without knowing f, g, or h. Nevertheless, measured data from the state and input trajectories are still required.

To demonstrate how learning-based PI works, notice that the system dynamics (4.32) can be expressed as

$$\begin{aligned}\dot{x} = & f(x) + g(x)u_i(x) + h(x)a_i(x) \\ & + g(x)(u - u_i(x)) + h(x)(a - a_i(x)),\end{aligned} \tag{4.40}$$

where the functions u_i and a_i are the same as those derived in the step $i-1 \in \mathbb{N}$ of Algorithm 4.4, and the argument of time has been dropped to simplify notation. Using (4.40), the total time derivative of the corresponding value function V_i along the trajectories of (4.32) is given by

$$\begin{aligned}\dot{V} = & \nabla V_i^{\mathrm{T}}(x)(f(x) + g(x)u_i(x) + h(x)a_i(x)) \\ & + \nabla V_i^{\mathrm{T}}(x)(g(x)(u - u_i(x)) + h(x)(a - a_i(x))),\end{aligned}$$

which, after exploiting (4.38) and (4.39), turns into

$$\begin{aligned}\dot{V} = & -Q(x) - u_i(x)^{\mathrm{T}} R u_i(x) + \gamma^2 a_i(x)^{\mathrm{T}} a_i(x) \\ & - 2u_{i+1}^{\mathrm{T}}(x) R(u - u_i(x)) + 2\gamma^2 a_{i+1}^{\mathrm{T}}(x)(a - a_i(x))).\end{aligned} \tag{4.41}$$

Consider now the time instants t_k, t'_k, $k \in \{0, 1, \ldots, K\} := \mathcal{K}$, such that $t_k \geq t_0$ and $t'_k > t_k$ for all $k \in \mathcal{K}$. Then integrating (4.41) over the time interval $[t_k, t'_k]$ yields

$$\begin{aligned}V_i(x(t'_k)) - V_i(x(t_k)) = & \int_{t_k}^{t'_k} \Big(- Q(x(\tau)) \\ & - u_i(x(\tau))^{\mathrm{T}} R u_i(x(\tau)) + \gamma^2 a_i(x(\tau))^{\mathrm{T}} a_i(x(\tau)) \\ & - 2u_{i+1}^{\mathrm{T}}(x(\tau)) R(u(\tau) - u_i(x(\tau))) \\ & + 2\gamma^2 a_{i+1}^{\mathrm{T}}(x(\tau))(a(\tau) - a_i(x(\tau)))\Big) \mathrm{d}\tau, \quad k \in \mathcal{K}.\end{aligned} \tag{4.42}$$

Eq. (4.42) provides a model-free method to express the value function V_i and the policies u_{i+1}, a_{i+1}; it expresses them as functions of measured state trajectories $x(t)$ and input trajectories $u(t)$ and $a(t)$ for $t \geq t_0$. This gives rise to the learning-based PI procedure, which is described in Algorithm 4.5. Effective methods to implement Algorithm 4.5 using actor-critic networks

Algorithm 4.5 Learning-Based PI.

Let $i = 0$, $\Omega \subset \mathbb{R}^n$, $\epsilon > 0$. Start with a tuple of policies $\{u_0, a_0\}$ that is stabilizing in Ω.
repeat
  Solve for V_i, u_{i+1}, and a_{i+1} over Ω from (4.42).
  Set $i = i + 1$.
until $i \geq 2$ & $\sup_{x \in \Omega} |V_{i-1}(x) - V_{i-2}(x)| < \epsilon$.

have been proposed, with convergence guarantees (Gao and Jiang, 2017; Jiang and Jiang, 2014; Jiang et al., 2020; Modares et al., 2015; Song et al., 2015).

4.3.4 Learning with clock offsets

It is evident that Algorithm 4.5 assumes perfect synchronization between all components of the CPS. As a consequence, it is not certain whether it will behave well, given even a relatively small clock offset; Algorithm 4.5 is, after all, a highly nonlinear process.

Motivated by the preceding, we will study whether the learning-based Algorithm 4.5 is robust in the epsilon–delta sense with respect to clock offsets in the components of the CPS. In particular, we will assume that the components of the CPS provide Algorithm 4.5 with the data (4.33), which have been corrupted by timing discrepancies. Consequently, at each time $t \geq t_0$, Algorithm 4.5 receives the following corrupted versions of $x(t)$, $u(t)$, and $a(t)$:

$$\bar{x}(t) = \begin{bmatrix} \bar{x}_1(t) \\ \bar{x}_2(t) \\ \vdots \\ \bar{x}_n(t) \end{bmatrix}, \quad \bar{u}(t) = \begin{bmatrix} \bar{u}_1(t) \\ \bar{u}_2(t) \\ \vdots \\ \bar{u}_m(t) \end{bmatrix}, \quad \bar{a}(t) = \begin{bmatrix} \bar{a}_1(t) \\ \bar{a}_2(t) \\ \vdots \\ \bar{a}_p(t) \end{bmatrix}.$$

Notice that the clock offsets that have corrupted the measured data are both unknown and time-varying, and hence it is not possible to cross them out. In fact, we may not even be aware of the existence of these offsets. As a result, it is not possible to construct Eq. (4.42) for Algorithm 4.5 and learn the functions V_i, u_{i+1}, and a_{i+1} directly; rather, we are forced to learn the functions $\bar{V}_i$, $\bar{u}_{i+1}$, and $\bar{a}_{i+1}$, which satisfy the clock-offseted version of (4.42):

$$
\begin{aligned}
\bar{V}_i(\bar{x}(t_k')) - V_i(\bar{x}(t_k)) = \int_{t_k}^{t_k'} \Big(& - Q(\bar{x}(\tau)) \\
& - \bar{u}_i(\bar{x}(\tau))^{\mathrm{T}} R \bar{u}_i(\bar{x}(\tau)) + \gamma^2 \bar{a}_i(\bar{x}(\tau))^{\mathrm{T}} \bar{a}_i(\bar{x}(\tau)) \\
& - 2\bar{u}_{i+1}^{\mathrm{T}}(\bar{x}(\tau)) R(\bar{u}(\tau) - \bar{u}_i(\bar{x}(\tau))) \\
& + 2\gamma^2 \bar{a}_{i+1}^{\mathrm{T}}(\bar{x}(\tau))(\bar{a}(\tau) - \bar{a}_i(\bar{x}(\tau))) \Big) \mathrm{d}\tau, \quad k \in \mathcal{K}.
\end{aligned} \tag{4.43}
$$

As a result of (4.43), the following question arises: will the "corrupted" value function $\bar{V}_i$ and the "corrupted" policies $\bar{u}_{i+1}$ and $\bar{a}_{i+1}$ converge close to the true ones? In the upcoming sections, we will prove that, under certain assumptions, the answer to this question is positive. In particular, we will show that if the clock offsets do not exceed a $\delta(\epsilon)$-threshold, and given certain continuity/differentiability assumptions, the functions $\bar{V}_i$, $\bar{u}_{i+1}$, and $\bar{a}_{i+1}$ enter an ϵ-neighborhood of $V^\star$, $u^\star$, and $a^\star$ over a compact set $\Omega \subset \mathbb{R}^n$ after a finite number of iterations in $i \in \mathbb{N}$.

Before we proceed to the main results, it is worth noting that we have implicitly assumed that the learning component of the CPS does not have its own clock; rather, it is synchronized with the reference/true clock, but without loss of generality. To see this, let $c_L : [t_0, \infty) \to [t_0, \infty)$ be the clock of the learning component. Then, instead of Eq. (4.43), we construct the equation

$$
\begin{aligned}
\bar{V}_i(\bar{x}(c_L(t_k'))) - V_i(\bar{x}(c_L(t_k))) = \int_{c_L(t_k)}^{c_L(t_k')} \Big(& - Q(\bar{x}(\tau)) \\
& - \bar{u}_i(\bar{x}(\tau))^{\mathrm{T}} R \bar{u}_i(\bar{x}(\tau)) + \gamma^2 \bar{a}_i(\bar{x}(\tau))^{\mathrm{T}} \bar{a}_i(\bar{x}(\tau)) \\
& - 2\bar{u}_{i+1}^{\mathrm{T}}(\bar{x}(\tau)) R(\bar{u}(\tau) - \bar{u}_i(\bar{x}(\tau))) \\
& + 2\gamma^2 \bar{a}_{i+1}^{\mathrm{T}}(\bar{x}(\tau))(\bar{a}(\tau) - \bar{a}_i(\bar{x}(\tau))) \Big) \mathrm{d}\tau, \quad k \in \mathcal{K},
\end{aligned} \tag{4.44}
$$

for $c_L(t_k') > c_L(t_k)$. Letting $\tilde{t}_k = c_L(t_k)$ and $\tilde{t}_k' = c_L(t_k')$, we can see that (4.44) is essentially identical to (4.43). The only difference is that the interval of integration is switched from $[t_k, t_k']$ to $[\tilde{t}_k, \tilde{t}_k']$, but the integration interval is allowed to be arbitrarily chosen anyway.

4.3.5 Learning-based PI with clock offsets

Since Eq. (4.43) is infinite-dimensional, it is difficult to solve explicitly for the value function V_i and the policies u_{i+1}, a_{i+1}, $i \in \mathbb{N}$. Nevertheless, these functions can be expressed as

$$
V_i(x) = (w_i^v)^{\mathrm{T}} \phi^v(x) + \epsilon_i^v(x),
$$

$$u_{i+1}(x) = (w_i^u)^{\mathrm{T}} \phi^u(x) + \epsilon_i^u(x),$$
$$a_{i+1}(x) = (w_i^a)^{\mathrm{T}} \phi^a(x) + \epsilon_i^a(x),$$

where $w_i^v \in \mathbb{R}^{N_v}$, $w_i^u \in \mathbb{R}^{N_u \times m}$, $w_i^a \in \mathbb{R}^{N_a \times p}$ are weight matrices, $\phi^v : \mathbb{R}^n \to \mathbb{R}^{N_v}$, $\phi^u : \mathbb{R}^n \to \mathbb{R}^{N_u}$, $\phi^a : \mathbb{R}^n \to \mathbb{R}^{N_a}$ are basis functions such that $\phi^v(0) = \phi^u(0) = \phi^a(0) = 0$, and $\epsilon_i^v : \mathbb{R}^n \to \mathbb{R}$, $\epsilon_i^u : \mathbb{R}^n \to \mathbb{R}^m$, $\epsilon_i^a : \mathbb{R}^n \to \mathbb{R}^p$ are approximation errors. It is known from the Weierstrass approximation theorem (Hornik et al., 1990) that the approximation errors ϵ_i^v, ϵ_i^u, and ϵ_i^a converge to zero, uniformly on any compact set $\Omega \subset \mathbb{R}^n$, as $N_v, N_u, N_a \to \infty$.

Still, the weight matrices w_i^v, w_i^u, w_i^a are not known in advance and must be identified through a learning-based procedure. For this purpose, an actor-critic network structure is employed, which approximates V_i, u_{i+1}, and a_{i+1} according to

$$\begin{aligned} \hat{V}_i(x) &= (\hat{w}_i^v)^{\mathrm{T}} \phi^v(x), \\ \hat{u}_{i+1}(x) &= (\hat{w}_i^u)^{\mathrm{T}} \phi^u(x), \\ \hat{a}_{i+1}(x) &= (\hat{w}_i^a)^{\mathrm{T}} \phi^a(x), \end{aligned} \tag{4.45}$$

where $\hat{w}_i^v \in \mathbb{R}^{N_v}$ are the critic weights, and $\hat{w}_i^u \in \mathbb{R}^{N_u \times m}$ and $\hat{w}_i^a \in \mathbb{R}^{N_a \times p}$ are the actor weights. The weights $\hat{w}_i^v$, $\hat{w}_i^u$, $\hat{w}_i^a$ then need to be trained through Eq. (4.42), so that $\hat{V}_i$, $\hat{u}_{i+1}$, and $\hat{a}_{i+1}$ become good approximations of V_i, u_{i+1}, and a_{i+1}. However, the issue here is that (4.42) cannot actually be constructed, owing to the clock offsets (4.34) that have corrupted the measured data. Hence we are forced to construct the offseted equation (4.43) instead and approximate the corrupted functions $\bar{V}_i$, $\bar{u}_{i+1}$, $\bar{a}_{i+1}$ in lieu of the actual ones.

Exploiting now the actor-critic structure (4.45), notice that the left-hand side of (4.43) can be approximated using the critic network, so that

$$\hat{V}_i(\bar{x}(t_k')) - \hat{V}_i(\bar{x}(t_k)) = (\phi^v(\bar{x}(t_k')) - \phi^v(\bar{x}(t_k)))^{\mathrm{T}} \hat{w}_i^v. \tag{4.46}$$

Additionally, the right-hand side of (4.43) can be approximated in terms of the offseted measured data, the actor policies at a previous step $i - 1 \in \mathbb{N}$, and the actor neural networks at step $i \in \mathbb{N}$. Specifically, the two right-most terms in (4.43) can be approximated in a recursive manner as

$$\begin{aligned} &2\hat{u}_{i+1}^{\mathrm{T}}(\bar{x}(\tau))R(\bar{u}(\tau) - \hat{u}_i(\bar{x}(\tau))) \\ &= 2(\phi^u(\bar{x}(\tau)))^{\mathrm{T}} \hat{w}_i^u R(\bar{u}(\tau) - \hat{u}_i(\bar{x}(\tau))) \\ &= 2\Big(\big((\bar{u}(\tau) - \hat{u}_i(\bar{x}(\tau)))^{\mathrm{T}} R\big) \otimes \phi^u(\bar{x}(\tau))^{\mathrm{T}}\Big) \mathrm{vec}(\hat{w}_i^u), \end{aligned} \tag{4.47}$$

$$\begin{aligned}
&2\gamma^2\hat{a}_{i+1}^{\mathrm{T}}(\bar{x}(\tau))(\bar{a}(\tau)-\hat{a}_i(\bar{x}(\tau)))\\
&\quad=2\gamma^2(\phi^a(\bar{x}(\tau)))^{\mathrm{T}}\hat{w}_i^a(\bar{a}(\tau)-\hat{a}_i(\bar{x}(\tau)))\\
&\quad=2\gamma^2\Big(\big((\bar{a}(\tau)-\hat{a}_i(\bar{x}(\tau)))^{\mathrm{T}}\big)\otimes\phi^a(\bar{x}(\tau))^{\mathrm{T}}\Big)\mathrm{vec}(\hat{w}_i^a).
\end{aligned}\tag{4.48}$$

The overall residual error by approximating Eq. (4.43) with the actor-critic structure is given, for $k\in\mathcal{K}$ and $i\in\mathbb{N}$, by

$$\begin{aligned}
e_{i,k}=\ &\hat{V}_i(\bar{x}(t_k'))-\hat{V}_i(\bar{x}(t_k))+\int_{t_k}^{t_k'}\Big(Q(\bar{x}(\tau))\\
&+\hat{u}_i(\bar{x}(\tau))^{\mathrm{T}}R\hat{u}_i(\bar{x}(\tau))+2\hat{u}_{i+1}^{\mathrm{T}}(\bar{x}(\tau))R(\bar{u}(\tau)-\hat{u}_i(\bar{x}(\tau)))\\
&-\gamma^2\hat{a}_i^{\mathrm{T}}(\bar{x}(\tau))\hat{a}_i(\bar{x}(\tau))-2\gamma^2\hat{a}_{i+1}^{\mathrm{T}}(\bar{x}(\tau))(\bar{a}(\tau)-\hat{a}_i(\bar{x}(\tau)))\Big)\mathrm{d}\tau.
\end{aligned}$$

Using (4.46)–(4.48), the residual error can be written in a linear form with respect to the actor-critic weights at step $i\in\mathbb{N}$ for all $k\in\mathcal{K}$, according to the formula

$$e_{i,k}=\bar{\Psi}_{i,k}\hat{W}_i+\bar{\Phi}_{i,k},\tag{4.49}$$

where $\hat{W}_i:=[(\hat{w}_i^v)^{\mathrm{T}}\ \mathrm{vec}(\hat{w}_i^u)^{\mathrm{T}}\ \mathrm{vec}(\hat{w}_i^a)^{\mathrm{T}}]^{\mathrm{T}}$, $\bar{\Psi}_{i,k}:=[\bar{\Psi}_{i,k}^v\ \bar{\Psi}_{i,k}^u\ \bar{\Psi}_{i,k}^a]$, and

$$\begin{aligned}
\bar{\Psi}_{i,k}^v&:=\Big(\phi^v(\bar{x}(t_k'))-\phi^v(\bar{x}(t_k))\Big)^{\mathrm{T}},\\
\bar{\Psi}_{i,k}^u&:=\int_{t_k}^{t_k'}2\big((\bar{u}(\tau)-\hat{u}_i(\bar{x}(\tau)))^{\mathrm{T}}R\big)\otimes\phi^u(\bar{x}(\tau))^{\mathrm{T}}\mathrm{d}\tau,\\
\bar{\Psi}_{i,k}^a&:=\int_{t_k}^{t_k'}-2\gamma^2\big((\bar{a}(\tau)-\hat{a}_i(\bar{x}(\tau)))^{\mathrm{T}}\big)\otimes\phi^a(\bar{x}(\tau))^{\mathrm{T}}\mathrm{d}\tau,\\
\bar{\Phi}_{i,k}&:=\int_{t_k}^{t_k'}\Big(Q(\bar{x}(\tau))+\hat{u}_i(\bar{x}(\tau))^{\mathrm{T}}R\hat{u}_i(\bar{x}(\tau))\\
&\qquad\qquad-\gamma^2\hat{a}_i(\bar{x}(\tau))^{\mathrm{T}}\hat{a}_i(\bar{x}(\tau))\Big)\mathrm{d}\tau.
\end{aligned}$$

This linear expression of the residual error with respect to the actor-critic weights allows us to use the least sum of squares as a training method. Naturally, then the following standard assumption (Jiang and Jiang, 2014; Modares et al., 2015) is required, which demands that the measured data contain sufficiently rich information about the system.

Algorithm 4.6 Learning-Based PI with Clock Offseted Data.

1: Let $i = 0$, $\Omega \subset \mathbb{R}^n$, $\epsilon > 0$. Start with a tuple of policies $\{\hat{u}_0, \hat{a}_0\} = \{u_0, a_0\}$ that is stabilizing in Ω.
2: **repeat**
3: Solve for $\hat{W}_i$ through (4.50).
4: Set $i = i + 1$.
5: **until** $i \geq 2$ & $\left\|\hat{W}_{i-1} - \hat{W}_{i-2}\right\| < \epsilon$.

Assumption 4.1. The measured data is sufficiently rich, i.e., there exist constants $\eta > 0$ and $K_0 \in \mathbb{N}$ such that if $K \geq K_0$, then

$$\frac{1}{K}\sum_{k=0}^{K} \bar{\Psi}_{i,k}^{\mathrm{T}} \bar{\Psi}_{i,k} \succ \eta I_{N_v+mN_u+pN_a}, \quad i \in \mathbb{N}. \quad \blacksquare$$

Given that Assumption 2.3 holds, the weights of the actor-critic structure can be trained at each step $i \in \mathbb{N}$ of the learning process according to the least-sum-of-squares law

$$\hat{W}_i = -\Big(\sum_{k=0}^{K} \bar{\Psi}_{i,k}^{\mathrm{T}} \bar{\Psi}_{i,k}\Big)^{-1} \Big(\sum_{k=0}^{K} \bar{\Psi}_{i,k}^{\mathrm{T}} \bar{\Phi}_{i,k}\Big). \tag{4.50}$$

The overall procedure learning procedure with clock offsets is described in Algorithm 4.6.

4.3.6 Convergence analysis

In what follows, we will show that the learning-based PI algorithm retains robustness in the epsilon–delta sense with respect to the clock offsets, which have corrupted the measured data. In particular, we will prove that $\hat{V}_i$, $\hat{u}_{i+1}$, and $\hat{a}_{i+1}$ will reach an ϵ-neighborhood of the optimal value function and the saddle-point policies, given that the magnitude of the offsets is uniformly confined below some $\delta(\epsilon)$.

To this end, for all $i \in \mathbb{N}$, let us consider the function $\tilde{V}_i$ that satisfies the Lyapunov equation

$$\begin{aligned}\nabla \tilde{V}_i^{\mathrm{T}}(x)(f(x) + g(x)\hat{u}_i(x) + h(x)\hat{a}_i(x)) \\ + Q(x) + \hat{u}_i(x)^{\mathrm{T}} R \hat{u}_i(x) - \gamma^2 \hat{a}_i(x)^{\mathrm{T}} \hat{a}_i(x) = 0.\end{aligned} \tag{4.51}$$

Additionally, let us define the following policy tuple as a function of $\tilde{V}_i$:

$$\begin{aligned}\tilde{u}_{i+1}(x) &= -\frac{1}{2}R^{-1}g^{\mathrm{T}}(x)\nabla\tilde{V}_i(x),\\ \tilde{a}_{i+1}(x) &= \frac{1}{2\gamma^2}h^{\mathrm{T}}(x)\nabla\tilde{V}_i(x).\end{aligned} \tag{4.52}$$

Following the same steps as in Section 4.3.3, by (4.51)–(4.52) it can be shown that over the trajectories of (4.32) we have

$$\begin{aligned}\tilde{V}_i(x(t'_k)) - \tilde{V}_i(x(t_k)) = \int_{t_k}^{t'_k} \Big(&- Q(x(\tau))\\ &- \hat{u}_i(x(\tau))^{\mathrm{T}}R\hat{u}_i(x(\tau)) + \gamma^2\hat{a}_i(x(\tau))^{\mathrm{T}}\hat{a}_i(x(\tau))\\ &- 2\tilde{u}_{i+1}^{\mathrm{T}}(x(\tau))R(u(\tau) - \hat{u}_i(x(\tau)))\\ &+ 2\gamma^2\tilde{a}_{i+1}^{\mathrm{T}}(x(\tau))(a(\tau) - \hat{a}_i(x(\tau)))\Big)\mathrm{d}\tau, \quad k \in \mathcal{K}.\end{aligned} \tag{4.53}$$

The following auxiliary lemma is the first step toward proving that Algorithm 4.6 converges close to the desired functions, given that the clock offsets are uniformly bounded below a certain threshold, provided that the number of basis functions is large enough. For its results to hold, the following additional assumption will be required.

Assumption 4.2. The following hold:

- The functions $\tilde{V}_i$ are continuously differentiable on Ω for all $i \in \mathbb{N}$.
- The trajectories of the state $x(t)$ are Lipschitz on t and confined in Ω for all $t \geq t_0$.
- The trajectories of the control inputs $u(t)$ and $a(t)$ are piecewise continuous on t and uniformly bounded for all $t \geq t_0$. ■

Define now the function $\Delta(t)$ to be, for all $t \geq t_0$, the greatest of all clock offsets' magnitudes:

$$\Delta(t) = \max\left\{\max_{i\in\mathcal{N}_x}|\delta_i^x(t)|,\ \max_{j\in\mathcal{N}_u}|\delta_j^u(t)|,\ \max_{l\in\mathcal{N}_a}|\delta_l^a(t)|\right\}.$$

Then we state the following auxiliary lemma.

Lemma 4.3. *Let Assumptions 4.1–4.2 hold, and consider the iteration provided by Algorithm 4.6 for all $i \in \mathbb{N}$. Then, for all $\epsilon > 0$, there exist constants $N_v^\star$, $N_u^\star$, $N_a^\star \in \mathbb{N}_+$ and a strictly positive clock offset upper bound $\Delta^\star > 0$ such* *that if* $N_v \geq N_v^\star$, $N_u \geq N_u^\star$, $N_a \geq N_a^\star$, *and* $\Delta(t) \leq \Delta^\star$ *for all* $t \geq t_0$, *then*

$$\begin{aligned}
&|\hat{V}_i(x) - \tilde{V}_i(x)| \leq \epsilon, \\
&\left\|\hat{u}_{i+1}(x) - \tilde{u}_{i+1}(x)\right\| \leq \epsilon, \\
&\left\|\hat{a}_{i+1}(x) - \tilde{a}_{i+1}(x)\right\| \leq \epsilon
\end{aligned}$$

for all $x \in \Omega$.

Proof. By the Weierstrass approximation theorem the functions $\tilde{V}_i$, $\tilde{u}_{i+1}$, $\tilde{a}_{i+1}$, $i \in \mathbb{N}$, can be uniformly approximated on Ω, so that

$$\begin{aligned}
\tilde{V}_i(x) &= (\tilde{w}_i^v)^{\mathrm{T}} \phi^v(x) + \tilde{\epsilon}_i^v(x), \\
\tilde{u}_{i+1}(x) &= (\tilde{w}_i^u)^{\mathrm{T}} \phi^u(x) + \tilde{\epsilon}_i^u(x), \\
\tilde{a}_{i+1}(x) &= (\tilde{w}_i^a)^{\mathrm{T}} \phi^a(x) + \tilde{\epsilon}_i^a(x).
\end{aligned} \tag{4.54}$$

The approximation errors $\tilde{\epsilon}_i^v : \mathbb{R}^n \to \mathbb{R}$, $\tilde{\epsilon}_i^u : \mathbb{R}^n \to \mathbb{R}^m$, $\tilde{\epsilon}_i^a : \mathbb{R}^n \to \mathbb{R}^p$ vanish uniformly on Ω as $N_v, N_u, N_a \to \infty$. Substituting (4.54) into (4.53), for $i \in \mathbb{N}$, we derive

$$0 = \Psi_{i,k} \tilde{W}_i + \Phi_{i,k} + \tilde{E}_{i,k}, \quad k \in \mathbb{N}, \tag{4.55}$$

where $\tilde{W}_i = [\tilde{w}_i^{v\mathrm{T}} \; \mathrm{vec}(\tilde{w}_i^u)^{\mathrm{T}} \; \mathrm{vec}(\tilde{w}_i^a)^{\mathrm{T}}]^{\mathrm{T}}$, $\Psi_{i,k} := [\Psi_{i,k}^v \; \Psi_{i,k}^u \; \Psi_{i,k}^a]$, and

$$\begin{aligned}
\Psi_{i,k}^v &= \left(\phi^v(x(t_k')) - \phi^v(x(t_k))\right)^{\mathrm{T}}, \\
\Psi_{i,k}^u &= \int_{t_k}^{t_k'} 2\left((u(\tau) - \hat{u}_i(x(\tau)))^{\mathrm{T}} R\right) \otimes \phi^u(x(\tau))^{\mathrm{T}} \mathrm{d}\tau, \\
\Psi_{i,k}^a &= \int_{t_k}^{t_k'} -2\gamma^2\left((a(\tau) - \hat{a}_i(x(\tau)))^{\mathrm{T}}\right) \otimes \phi^a(x(\tau))^{\mathrm{T}} \mathrm{d}\tau, \\
\Phi_{i,k} &= \int_{t_k}^{t_k'} \Big(Q(x(\tau)) + \hat{u}_i(x(\tau))^{\mathrm{T}} R \hat{u}_i(x(\tau)) \\
&\qquad\qquad - \gamma^2 \hat{a}_i(x(\tau))^{\mathrm{T}} \hat{a}_i(x(\tau))\Big) \mathrm{d}\tau, \\
\tilde{E}_{i,k} &= \tilde{\epsilon}_i^v(x(t_k')) - \tilde{\epsilon}_i^v(x(t_k)) \\
&\quad + \int_{t_k}^{t_k'} 2\tilde{\epsilon}_i^u(x(\tau))^{\mathrm{T}} R(u(\tau) - \hat{u}_i(x(\tau))) \mathrm{d}\tau \\
&\quad - \int_{t_k}^{t_k'} 2\gamma^2 \tilde{\epsilon}_i^a(x(\tau))^{\mathrm{T}} (a(\tau) - \hat{a}_i(x(\tau))) \mathrm{d}\tau.
\end{aligned}$$

Adding and subtracting identical terms in (4.55), we have that for all $k \in \mathcal{K}$ and $i \in \mathbb{N}$,

$$0 = \bar{\Psi}_{i,k}\tilde{W}_i + \bar{\Phi}_{i,k} + \Delta\Psi_{i,k}\tilde{W}_i + \Delta\Phi_{i,k} + \tilde{E}_{i,k}, \tag{4.56}$$

where $\Delta\Psi_{i,k} = \Psi_{i,k} - \bar{\Psi}_{i,k}$ and $\Delta\Phi_{i,k} = \Phi_{i,k} - \bar{\Phi}_{i,k}$.

Since $\hat{W}_i$ is estimated through the least-sum-of-squares law (4.50) to minimize the sum of squares of the errors in (4.49), by Assumption 4.1 and (4.56) we have

$$\sum_{k=0}^{K} e_{i,k}^2 \le \sum_{k=0}^{K} (\tilde{E}_{i,k} + \Delta\Psi_{i,k}\tilde{W}_i + \Delta\Phi_{i,k})^2, \ i \in \mathbb{N}. \tag{4.57}$$

Subtracting (4.56) from (4.49), we obtain

$$e_{i,k} + \tilde{E}_{i,k} + \Delta\Psi_{i,k}\tilde{W}_i + \Delta\Phi_{i,k} = \bar{\Psi}_{i,k}(\hat{W}_i - \tilde{W}_i). \tag{4.58}$$

Multiplying (4.58) by itself and summing over k leads, due to Assumption 4.1, to

$$\begin{aligned} &\frac{1}{K}\sum_{k=0}^{K}(e_{i,k} + \tilde{E}_{i,k} + \Delta\Psi_{i,k}\tilde{W}_i + \Delta\Phi_{i,k})^2 \\ =&\frac{1}{K}\sum_{k=0}^{K}(\hat{W}_i - \tilde{W}_i)^{\mathrm{T}}\bar{\Psi}_{i,k}^{\mathrm{T}}\bar{\Psi}_{i,k}(\hat{W}_i - \tilde{W}_i) \ge \eta \left\| \hat{W}_i - \tilde{W}_i \right\|^2. \end{aligned} \tag{4.59}$$

However, by (4.57)

$$\begin{aligned} &\frac{1}{K}\sum_{k=0}^{K}(e_{i,k} + \tilde{E}_{i,k} + \Delta\Psi_{i,k}\tilde{W}_i + \Delta\Phi_{i,k})^2 \\ &\quad \le \frac{4}{K}\sum_{k=0}^{K}(\tilde{E}_{i,k} + \Delta\Psi_{i,k}\tilde{W}_i + \Delta\Phi_{i,k})^2 \\ &\quad \le \max_{1\le k\le K} 4(\tilde{E}_{i,k} + \Delta\Psi_{i,k}\tilde{W}_i + \Delta\Phi_{i,k})^2, \end{aligned}$$

which, combined with (4.59), yields

$$\max_{1\le k\le K} \frac{4}{\eta}(\tilde{E}_{i,k} + \Delta\Psi_{i,k}\tilde{W}_i + \Delta\Phi_{i,k})^2 \ge \left\| \hat{W}_i - \tilde{W}_i \right\|^2. \tag{4.60}$$

Now notice that by Assumption 4.2, $\tilde{E}_{i,k} \to 0$ as $N_v, N_u, N_a \to \infty$ (uniformly on any trajectories of x on Ω). Additionally, we have

$$\begin{aligned}
\Delta\Psi_{i,k}\tilde{W}_i = \int_{t_k}^{t'_k} \Big(&2(\tilde{u}_{i+1}(x(\tau)) - \tilde{\epsilon}_i^u(x(\tau))^{\mathrm{T}} R(u(\tau) - \hat{u}_i(x(\tau))) \\
&- 2(\tilde{u}_{i+1}(\bar{x}(\tau)) - \tilde{\epsilon}_i^u(\bar{x}(\tau))^{\mathrm{T}} R(\bar{u}(\tau) - \hat{u}_i(\bar{x}(\tau))) \\
&- 2\gamma^2(\tilde{a}_{i+1}(x(\tau)) - \tilde{\epsilon}_i^a(x(\tau)))^{\mathrm{T}}(a(\tau) - \hat{a}_i(x(\tau))) \\
&+ 2\gamma^2(\tilde{a}_{i+1}(\bar{x}(\tau)) - \tilde{\epsilon}_i^a(\bar{x}(\tau)))^{\mathrm{T}}(\bar{a}(\tau) - \hat{a}_i(\bar{x}(\tau)))\Big)\mathrm{d}\tau, \\
\Delta\Phi_{i,k} = \tilde{V}_i(x(t'_k)) &- \tilde{V}_i(x(t_k)) - \tilde{V}_i(\bar{x}(t'_k)) + \tilde{V}_i(\bar{x}(t_k)).
\end{aligned}$$

Using Assumption 4.2, we have

$$\begin{aligned}
|\Delta\Phi_{i,k}| &\le |\tilde{V}_i(x(t'_k)) - \tilde{V}_i(\bar{x}(t'_k))| + |\tilde{V}_i(x(t_k)) - \tilde{V}_i(\bar{x}(t_k))| \\
&\le L_{V_i}\left\|x(t'_k) - \bar{x}(t'_k)\right\| + L_{V_i}\|x(t_k) - \bar{x}(t_k)\| \\
&\le L_{V_i}\sum_{i=1}^{n}\left(|x_i(t'_k) - \bar{x}_i(t'_k)| + |x_i(t_k) - \bar{x}_i(t_k)|\right) \\
&= L_{V_i}\sum_{i=1}^{n}\Big(|x_i(t'_k) - x_i(t'_k + \delta_i^x(t'_k))| \\
&\qquad\qquad + |x_i(t_k) - x_i(t_k + \delta_i^x(t_k))|\Big) \\
&\le L_{V_i}L_x\sum_{i=1}^{n}(|\delta_i^x(t_k)| + |\delta_i^x(t'_k)|),
\end{aligned}$$

where L_{V_i} is the Lipschitz constant of $\tilde{V}_i$ on Ω, which exists since $\tilde{V}_i$ is continuously differentiable, and L_x is the Lipschitz constant of $x(t)$ on t. Hence, if $\Delta(t) \le \Delta_M$, then

$$|\Delta\Phi_{i,k}| \le 2nL_{V_i}L_x\Delta_M. \tag{4.61}$$

Inequality (4.61) implies that $\Delta\Phi_{i,k} \to 0$ (uniformly on any trajectories of x on Ω) as $\Delta_M \searrow 0$ and $N_v, N_u, N_a \to \infty$. Using Assumption 4.2, the same property can be similarly proved for $\Delta\Psi_{i,k}$. Therefore by (4.60), for every ϵ_1, there exist constants N_v^m, N_u^m, N_a^m, and an upper clock mismatch bound Δ^m such that if $N_v \ge N_v^m$, $N_u \ge N_u^m$, $N_a \ge N_a^m$, and $\Delta(t) \le \Delta^m$ for all $t \ge t_0$, then $\left\|\hat{W}_i - \tilde{W}_i\right\| \le \epsilon_1$. Thus, for all $\epsilon > 0$, there exist $N_v^\star$, $N_u^\star$, $N_a^\star > 0$ and a clock offset upper bound $\Delta^\star > 0$ such that if $N_v \ge N_v^\star$, $N_u \ge N_u^\star$, $N_a \ge N_a^\star$, and $\|\Delta(t)\| \leq \Delta^\star$ for all $t \geq t_0$, then

$$|\hat{V}_i(x) - \tilde{V}_i(x)| \leq \left\|(\hat{w}_i^v - \tilde{w}_i^v)\right\| \left\|\phi_i^v(x)\right\| + |\tilde{\epsilon}_i^v(x)| \leq \frac{\epsilon}{2} + \frac{\epsilon}{2} = \epsilon,$$

$$\left\|\hat{u}_{i+1}(x) - \tilde{u}_{i+1}(x)\right\| \leq \left\|\hat{w}_i^u - \tilde{w}_i^u\right\| \left\|\phi_i^u(x)\right\| + \left\|\tilde{\epsilon}_i^u(x)\right\| \leq \frac{\epsilon}{2} + \frac{\epsilon}{2} = \epsilon,$$

$$\left\|\hat{a}_{i+1}(x) - \tilde{a}_{i+1}(x)\right\| \leq \left\|\hat{w}_i^a - \tilde{w}_i^a\right\| \left\|\phi_i^a(x)\right\| + \left\|\tilde{\epsilon}_i^a(x)\right\| \leq \frac{\epsilon}{2} + \frac{\epsilon}{2} = \epsilon$$

for all $x \in \Omega$, which is the required result. □

The following theorem generalizes Lemma 4.3 and proves the desired epsilon–delta robustness result that we sought for the learning-based PI procedure. In particular, it shows that Algorithm 4.6 will eventually converge to an ϵ-suboptimal tuple of policies, given that the magnitude of the clock offsets remains below a threshold dependent on ϵ, and the number of actor-critic nodes is sufficient.

Theorem 4.3. *Let Assumptions 4.1–4.2 hold, and consider the procedure provided by Algorithm 4.6 for all $i \in \mathbb{N}$. Then, for all $\epsilon > 0$, there exist constants $N_v^{\star\star}$, $N_u^{\star\star}$, $N_a^{\star\star}$, $i^\star \in \mathbb{N}_+$, and an upper clock mismatch bound $\Delta^{\star\star} > 0$ such that if $N_v \geq N_v^{\star\star}$, $N_u \geq N_u^{\star\star}$, $N_a \geq N_a^{\star\star}$, and $\Delta(t) \leq \Delta^{\star\star}$ for all $t \geq t_0$, then*

$$\left\|\hat{V}_{i^\star}(x) - V^\star(x)\right\| \leq \epsilon,$$
$$\left\|\hat{u}_{i^\star+1}(x) - u^\star(x)\right\| \leq \epsilon,$$
$$\left\|\hat{a}_{i^\star+1}(x) - a^\star(x)\right\| \leq \epsilon$$

for all $x \in \Omega$.

Proof. Let us suppose that for some $\bar{\Delta} > 0$, $\Delta(t) \leq \bar{\Delta}$ for all $t \geq t_0$. The proof follows an inductive procedure.

For $i = 0$, we have $\tilde{V}_0 = V_0$, $\tilde{u}_1 = u_1$, and $\tilde{a}_1 = a_1$, since $\hat{u}_0 = u_0$ and $\hat{a}_0 = a_0$. Hence it follows from Lemma 4.3 that $\lim_{N_v,N_u,N_a\to\infty,\bar{\Delta}\searrow 0}\{\hat{V}_0(x), \hat{u}_1(x), \hat{a}_1(x)\} = \{V_0(x), u_1(x), a_1(x)\}$, uniformly on Ω.

Consider now the inductive assumption that $\lim_{N_v,N_u,N_a\to\infty,\bar{\Delta}\searrow 0}\{\hat{V}_{i-1}(x), \hat{u}_i(x), \hat{a}_i(x)\} = \{V_{i-1}(x), u_i(x), a_i(x)\}$, uniformly on Ω, for some $i \in \mathbb{N}_+$. Then, by the definitions of V_i and $\tilde{V}_i$, over the trajectories of any inputs

$u: \mathbb{R}^n \to \mathbb{R}^m$ and $a: \mathbb{R}^n \to \mathbb{R}^p$ satisfying Assumption 4.2, we have

$$
\begin{aligned}
&\left|V_i(x(t)) - \tilde{V}_i(x(t))\right| = \\
&\left|\int_t^\infty \Big(u_i(x(\tau))^{\mathrm{T}} R u_i(x(\tau)) + 2u_{i+1}^{\mathrm{T}}(x(\tau)) R v_i^u(x(\tau))\Big)\mathrm{d}\tau\right. \\
&- \int_t^\infty \Big(\hat{u}_i(x(\tau))^{\mathrm{T}} R \hat{u}_i(x(\tau)) + 2\tilde{u}_{i+1}^{\mathrm{T}}(x(\tau)) R \hat{v}_i^u(x(\tau))\Big)\mathrm{d}\tau \\
&- \int_t^\infty \Big(\gamma^2 a_i(x(\tau))^{\mathrm{T}} a_i(x(\tau)) + 2\gamma^2 a_{i+1}^{\mathrm{T}}(x(\tau)) v_i^a(x(\tau))\Big)\mathrm{d}\tau \\
&\left.+ \int_t^\infty \Big(\gamma^2 \hat{a}_i(x(\tau))^{\mathrm{T}} \hat{a}_i(x(\tau)) + 2\gamma^2 \tilde{a}_{i+1}^{\mathrm{T}}(x(\tau)) \hat{v}_i^a(x(\tau))\Big)\mathrm{d}\tau\right| \\
&\leq \left|\int_t^\infty \Big(u_i(x(\tau))^{\mathrm{T}} R u_i(x(\tau)) - \hat{u}_i(x(\tau))^{\mathrm{T}} R \hat{u}_i(x(\tau))\Big)\mathrm{d}\tau\right| \\
&+ \left|\int_t^\infty 2\Big((u_{i+1}(x(\tau)) - \tilde{u}_{i+1}(x(\tau)))^{\mathrm{T}} R \hat{v}_i^u(x(\tau))\Big)\mathrm{d}\tau\right| \\
&+ \left|\int_t^\infty \Big(2u_{i+1}^{\mathrm{T}}(x(\tau)) R(\hat{u}_i(x(\tau)) - u_i(x(\tau))\Big)\mathrm{d}\tau\right| \\
&+ \left|\int_t^\infty \Big(\gamma^2 a_i(x(\tau))^{\mathrm{T}} a_i(x(\tau)) - \gamma^2 \hat{a}_i(x(\tau))^{\mathrm{T}} \hat{a}_i(x(\tau))\Big)\mathrm{d}\tau\right| \\
&+ \left|\int_t^\infty 2\gamma^2\Big((a_{i+1}(x(\tau)) - \tilde{a}_{i+1}(x(\tau)))^{\mathrm{T}} \hat{v}_i^a(x(\tau))\Big)\mathrm{d}\tau\right| \\
&+ \left|\int_t^\infty \Big(2\gamma^2 a_{i+1}^{\mathrm{T}}(x(\tau))(\hat{a}_i(x(\tau)) - a_i(x(\tau))\Big)\mathrm{d}\tau\right|,
\end{aligned}
\tag{4.62}
$$

where $v_i^u = u - u_i$, $\hat{v}_i^u = u - \hat{u}_i$, $v_i^a = a - a_i$, and $\hat{v}_i^a = a - \hat{a}_i$. By the assumptions of the induction we have

$$
\begin{aligned}
&\lim_{\substack{N_u, N_v, N_a \to \infty \\ \bar{\Delta} \searrow 0}} \int_t^\infty \Big(u_i(x(\tau))^{\mathrm{T}} R u_i(x(\tau)) - \hat{u}_i(x(\tau))^{\mathrm{T}} R \hat{u}_i(x(\tau))\Big)\mathrm{d}\tau = 0, \\
&\lim_{\substack{N_u, N_v, N_a \to \infty \\ \bar{\Delta} \searrow 0}} \int_t^\infty 2u_{i+1}^{\mathrm{T}}(x(\tau)) R(\hat{u}_i(x(\tau)) - u_i(x(\tau)))\mathrm{d}\tau = 0, \\
&\lim_{\substack{N_u, N_v, N_a \to \infty \\ \bar{\Delta} \searrow 0}} \int_t^\infty \Big(\gamma^2 a_i(x(\tau))^{\mathrm{T}} a_i(x(\tau)) - \gamma^2 \hat{a}_i(x(\tau))^{\mathrm{T}} R \hat{a}_i(x(\tau))\Big)\mathrm{d}\tau = 0,
\end{aligned}
\tag{4.63}
$$

$$
\lim_{\substack{N_u, N_v, N_a \to \infty \\ \bar{\Delta} \searrow 0}} \int_t^\infty 2\gamma^2 a_{i+1}^{\mathrm{T}}(x(\tau))(\hat{a}_i(x(\tau)) - a_i(x(\tau)))\mathrm{d}\tau = 0.
$$

In addition, by Assumption 4.1, the definitions of $\tilde{V}_i$, $\tilde{u}_{i+1}$, $\tilde{a}_{i+1}$, and the inductive assumption we have on Ω that

$$\begin{aligned} \lim_{\substack{N_u,N_v,N_a\to\infty\\ \bar{\Delta}\searrow 0}} \tilde{u}_{i+1}(x) &= u_{i+1}(x), \\ \lim_{\substack{N_u,N_v,N_a\to\infty\\ \bar{\Delta}\searrow 0}} \tilde{a}_{i+1}(x) &= a_{i+1}(x). \end{aligned} \tag{4.64}$$

Hence, from (4.62), (4.63), and (4.64) we conclude that $\lim_{N_u,N_v,N_a\to\infty,\bar{\Delta}\searrow 0} \tilde{V}_i(x) = V_i(x)$ uniformly on Ω. On the other hand, from Lemma 4.3 we have

$$\lim_{N_u,N_v,N_a\to\infty,\bar{\Delta}\searrow 0} \{\hat{V}_i(x), \hat{u}_{i+1}(x), \hat{a}_{i+1}(x)\} = \{\tilde{V}_i(x), \tilde{u}_{i+1}(x), \tilde{a}_{i+1}(x)\}$$

uniformly on Ω. Combining the two results, we conclude that for all $\epsilon_2 > 0$, there exist constants $N_v^{\star\star}$, $N_u^{\star\star}$, $N_a^{\star\star}$ and a clock mismatch upper bound $\Delta^{\star\star}$ such that if $N_v \geq N_v^{\star\star}$, $N_u \geq N_u^{\star\star}$, $N_a \geq N_a^{\star\star}$, and $\Delta(t) \leq \Delta^{\star\star}$ for all $t \geq t_0$, then

$$\begin{aligned} \left\| \hat{V}_i(x) - V_i(x) \right\| &\leq \left\| \hat{V}_i(x) - \tilde{V}_i(x) \right\| + \left\| \tilde{V}_i(x) - V_i(x) \right\| \\ &\leq \frac{\epsilon_2}{2} + \frac{\epsilon_2}{2} = \epsilon_2, \\ \left\| \hat{u}_{i+1}(x) - u_{i+1}(x) \right\| &\leq \left\| \hat{u}_{i+1}(x) - \tilde{u}_{i+1}(x) \right\| \\ &\quad + \|\tilde{u}_{i+1}(x) - u_{i+1}(x)\| \leq \frac{\epsilon_2}{2} + \frac{\epsilon_2}{2} = \epsilon_2, \\ \left\| \hat{a}_{i+1}(x) - a_{i+1}(x) \right\| &\leq \left\| \hat{a}_{i+1}(x) - \tilde{a}_{i+1}(x) \right\| \\ &\quad + \|\tilde{a}_{i+1}(x) - a_{i+1}(x)\| \leq \frac{\epsilon_2}{2} + \frac{\epsilon_2}{2} = \epsilon_2, \end{aligned} \tag{4.65}$$

which completes the induction. Finally, from Wu and Luo (2012) we have that for every $\epsilon_3 > 0$, there exists $i^\star$ such that for all $i \geq i^\star$ and $x \in \Omega$,

$$\begin{aligned} &\left\| V^\star(x) - V_i(x) \right\| \leq \epsilon_3, \\ &\left\| u^\star(x) - u_{i+1}(x) \right\| \leq \epsilon_3, \quad \left\| a^\star(x) - a_{i+1}(x) \right\| \leq \epsilon_3. \end{aligned} \tag{4.66}$$

Hence the proof follows from (4.65) and (4.66) using the triangle equality. □

4.3.7 Learning-based PI with clock offsets and quantization

We now consider that measurements (4.33) of the state do not only suffer from clock offsets, but also from quantization. Specifically, we employ the logarithmic quantizer considered in Elia and Mitter (2001); Okano et al. (2017): given $\alpha_0 > 0$ and $\rho \in (0,\ 1)$, the quantizer $r : \mathbb{R} \to \mathbb{R}$ is defined as

$$r(y) = \begin{cases} \frac{2\rho}{\rho+1}\rho^\ell \alpha_0 & \text{if } y \in (\rho^{\ell+1}\alpha_0,\ \rho^\ell \alpha_0], \\ 0 & \text{if } y = 0, \\ -\frac{2\rho}{\rho+1}\rho^\ell \alpha_0 & \text{if } y \in [-\rho^\ell \alpha_0,\ -\rho^{\ell+1}\alpha_0), \end{cases} \tag{4.67}$$

where $\ell \in \mathbb{Z}$ is the integer such that $\rho^{\ell+1}\alpha_0 < |y| \leq \rho^\ell \alpha_0$. Evidently, the quantization levels become finer as $\rho \to 1$ and coarser as $\rho \to 0$. It is also helpful to define as $r^n : \mathbb{R}^n \to \mathbb{R}^n$ the quantizer that applies the operator (4.67) entrywise on vectors in $\mathbb{R}^n$.

The quantization further dilutes the measurements of the state signals; instead of measuring $\bar{x}_i(t)$, $i \in \mathcal{N}_x$, as in (4.33), we now only have access to the quantized values of $\bar{x}_i(t)$:

$$\hat{x}_i(t) = r(\bar{x}_i(t)) = r(x_i(c_i^x(t))) = r(x_i(t + \delta_i^x(t)))$$

with the whole quantized vector

$$\hat{x}(t) = \begin{bmatrix} \hat{x}_1(t) & \hat{x}_2(t) & \dots & \hat{x}_n(t) \end{bmatrix}^{\mathrm{T}}.$$

Notice that a quantization of u or a will not have any impact on the learning-based PI, as both measured and actual values of the control input signals implemented in (4.32) will be quantized. On the other hand, although the measured value of the state x is quantized, its actual value is continuous owing to the continuous-flow nature of (4.32). As such, discrepancies will exist between state measurements and reality, which could jeopardize the convergence of the learning-based PI.

Toward analyzing the learning-based PI algorithm under the logarithmic quantization, we need the following lemma.

Lemma 4.4. *Let $z \in D \subset \mathbb{R}^n$, where D is a compact set. Then the quantization error $\|r^n(z) - z\|$ is uniformly bounded on D. In addition, $\lim_{\rho \nearrow 1} \|r^n(z) - z\| = 0$ uniformly on D.*

Proof. The uniform boundedness of the quantization error on D follows immediately from the uniform boundedness of both z and $r^n(z)$ on D by definition. To prove the uniform convergence, let z_i be the ith entry of z, $i \in \mathcal{N}_x$. Then there are three cases:

i) If $z_i = 0$, then $|r(z_i) - z_i| = 0$ by definition.

ii) Let $z_i > 0$. Then $r(z_i) = \frac{2\rho}{\rho+1}\rho^\ell \alpha_0$, where $\ell \in \mathbb{Z}$ is such that $\rho^{\ell+1}\alpha_0 < z_i \leq \rho^\ell \alpha_0$. Consequently,

$$\rho^\ell \alpha_0 \left(\frac{2\rho}{\rho+1} - 1\right) \leq r(z_i) - z_i < \rho^\ell \alpha_0 \left(\frac{2\rho}{\rho+1} - \rho\right),$$

which yields

$$|r(z_i) - z_i| < \rho^\ell \alpha_0 \left(\frac{2\rho}{\rho+1} - \rho\right). \tag{4.68}$$

Nevertheless, $\rho^\ell \alpha_0 = \rho^{-1}\rho^{\ell+1}\alpha_0 < \rho^{-1}z_i \leq \rho^{-1}z_m$, where $z_m = \max_{z\in D}\|z\|_\infty$, and the maximum exists since D is compact. Therefore (4.68) becomes

$$|r(z_i) - z_i| < z_m \left(\frac{2}{\rho+1} - 1\right).$$

It is now evident that for all $\epsilon > 0$ such that $|r(z_i) - z_i| < \epsilon$, there exists $\rho^\star \in [0, 1)$, where $\rho^\star = \max\{\frac{2z_m}{z_m+\epsilon} - 1, 0\}$, such that if $\rho \in (\rho^\star, 1)$, then $|r(z_i) - z_i| < \epsilon$. Hence the uniform convergence is proved.

iii) The case where $z_i < 0$ is similar to ii). □

Exploiting Lemma 4.4, the analysis of the previous section can be extended to the case of quantization. Particularly, we can show that the learning-based PI algorithm always converges in an ϵ neighborhood of the optimal control and value function, given a sufficiently fine quantizer. The actual equations of the learning-based PI with quantization will be identical to the quantization-free case, with the exception that the quantized data are utilized instead. Particularly, the weights are trained as

$$\hat{W}_i = -\Big(\sum_{k=0}^{K} \hat{\Psi}_{i,k}^{\mathrm{T}} \hat{\Psi}_{i,k}\Big)^{-1} \Big(\sum_{k=0}^{K} \hat{\Psi}_{i,k}^{\mathrm{T}} \hat{\Phi}_{i,k}\Big), \tag{4.69}$$

where $\hat{W}_i := [(\hat{w}_i^v)^{\mathrm{T}}\ \mathrm{vec}(\hat{w}_i^u)^{\mathrm{T}}\ \mathrm{vec}(\hat{w}_i^a)^{\mathrm{T}}]^{\mathrm{T}}$, $\hat{\Psi}_{i,k} := [\hat{\Psi}_{i,k}^v\ \hat{\Psi}_{i,k}^u\ \hat{\Psi}_{i,k}^a]$, and

$$\hat{\Psi}_{i,k}^v := \Big(\phi^v(\hat{x}(t_k')) - \phi^v(\hat{x}(t_k))\Big)^{\mathrm{T}},$$

$$\hat{\Psi}_{i,k}^u := \int_{t_k}^{t_k'} 2\big((\bar{u}(\tau) - \hat{u}_i(\hat{x}(\tau)))^{\mathrm{T}} R\big) \otimes \phi^u(\hat{x}(\tau))^{\mathrm{T}} \mathrm{d}\tau,$$

$$\hat{\Psi}_{i,k}^a := \int_{t_k}^{t_k'} -2\gamma^2\big((\bar{a}(\tau) - \hat{a}_i(\hat{x}(\tau)))^{\mathrm{T}}\big) \otimes \phi^a(\hat{x}(\tau))^{\mathrm{T}} \mathrm{d}\tau,$$

Algorithm 4.7 Learning-Based PI with Clock Offseted and Quantized Data.

1: Let $i = 0$, $\Omega \subset \mathbb{R}^n$, $\epsilon > 0$. Start with a tuple of policies $\{\hat{u}_0, \hat{a}_0\} = \{u_0, a_0\}$ that is stabilizing in Ω.
2: **repeat**
3: Solve for $\hat{W}_i$ through (4.69).
4: Set $i = i + 1$.
5: **until** $i \geq 2$ & $\left\| \hat{W}_{i-1} - \hat{W}_{i-2} \right\| < \epsilon$.

$$\hat{\Phi}_{i,k} := \int_{t_k}^{t'_k} \Big(Q(\hat{x}(\tau)) + \hat{u}_i(\hat{x}(\tau))^{\mathrm{T}} R \hat{u}_i(\hat{x}(\tau)) - \gamma^2 \hat{a}_i(\hat{x}(\tau))^{\mathrm{T}} \hat{a}_i(\hat{x}(\tau)) \Big) \mathrm{d}\tau.$$

The corresponding procedure is shown in Algorithm 4.7.

For (4.69) to be properly defined, we need a stronger form of Assumption 4.1, requiring the quantized version of the measured data to be sufficiently rich.

Assumption 4.3. The quantized measured data is sufficiently rich, i.e., there exist constants $\hat{\eta} > 0$ and $\hat{K}_0 \in \mathbb{N}$ such that if $\hat{K} \geq K_0$, then

$$\frac{1}{\hat{K}} \sum_{k=0}^{\hat{K}} \hat{\Psi}_{i,k}^{\mathrm{T}} \hat{\Psi}_{i,k} \succ \hat{\eta} I_{N_v + mN_u + pN_a}, \quad i \in \mathbb{N}. \quad \blacksquare$$

The convergence properties of learning-based PI under both clock offsets and quantization are summarized in the following theorem.

Theorem 4.4. *Let Assumptions 4.2–4.3 hold, and consider the procedure provided by Algorithm 4.7 for all $i \in \mathbb{N}$. Then, for all $\epsilon > 0$, there exist constants $N_v^{\star\star\star}$, $N_u^{\star\star\star}$, $N_a^{\star\star\star}$, $i^\star \in \mathbb{N}_+$, an upper clock mismatch bound $\Delta^{\star\star\star} > 0$, and a low bound $\rho^\star \in [0, 1)$ for the quantizer coarseness, such that if $N_v \geq N_v^{\star\star\star}$, $N_u \geq N_u^{\star\star\star}$, $N_a \geq N_a^{\star\star\star}$, $\rho \in (\rho^\star, 1)$, and $\Delta(t) \leq \Delta^{\star\star\star}$ for all $t \geq t_0$, then*

$$\left\| \hat{V}_{i^\star}(x) - V^\star(x) \right\| \leq \epsilon,$$
$$\left\| \hat{u}_{i^\star+1}(x) - u^\star(x) \right\| \leq \epsilon,$$
$$\left\| \hat{a}_{i^\star+1}(x) - a^\star(x) \right\| \leq \epsilon$$

for all $x \in \Omega$.

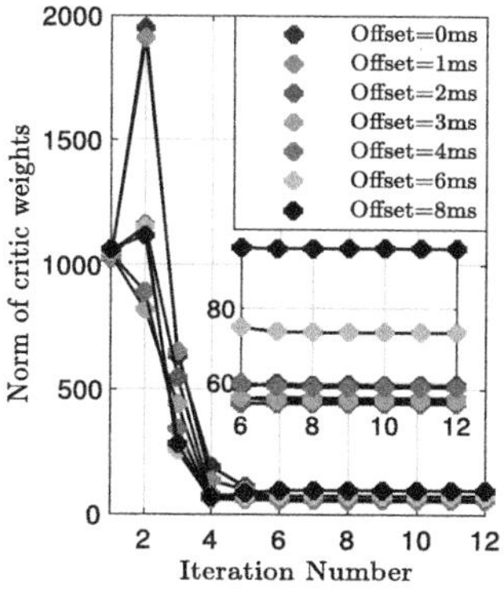

(a) Evolution of the norm of the critic weights at each iteration of the learning-based PI with clock mismatches.

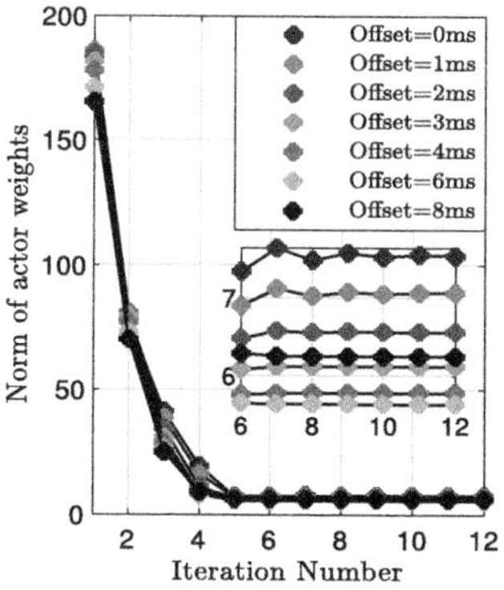

(b) Evolution of the norm of the actor weights at each iteration of the learning-based PI with clock mismatches.

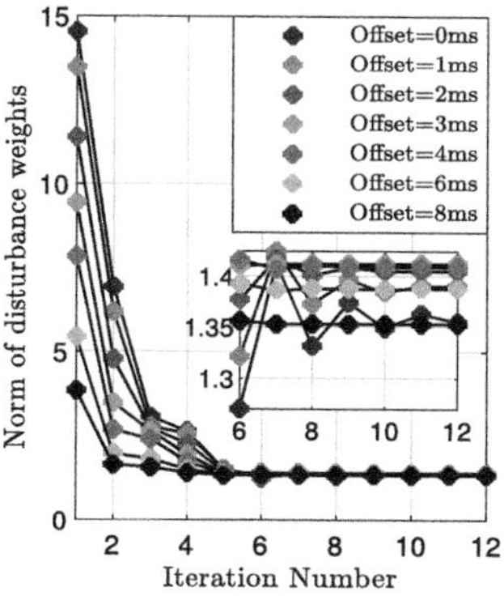

(c) Evolution of the norm of the disturbance weights at each iteration of the learning-based PI with clock mismatches.

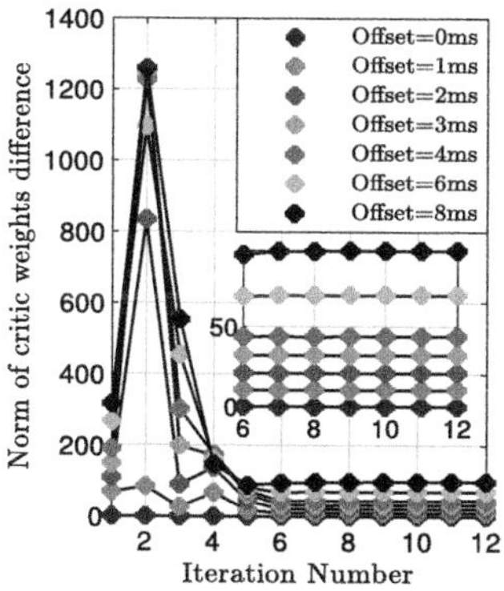

(d) Evolution of the norm of the difference of the critic weights at each scenario with the critic weights of the offset-free scenario.

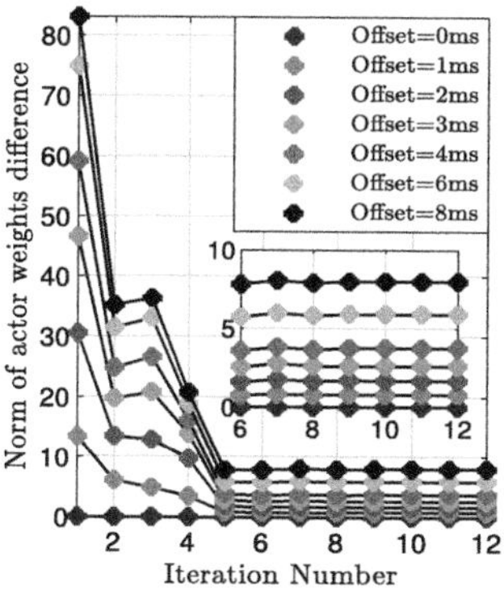

(e) Evolution of the norm of the difference of the actor weights at each scenario with the actor weights of the offset-free scenario.

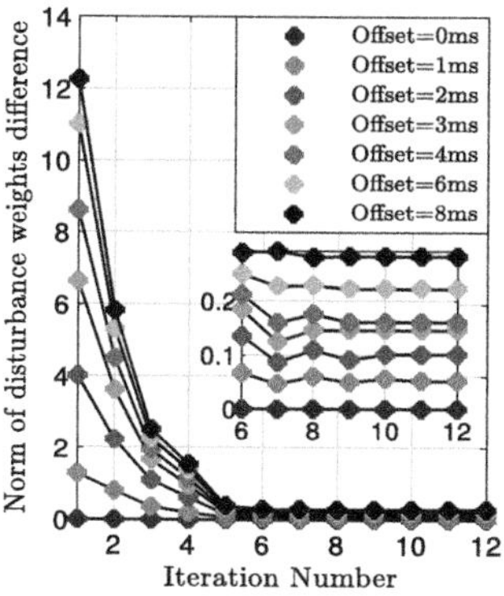

(f) Evolution of the norm of the difference of the disturbance weights at each scenario with the disturbance weights of the offset-free scenario.

Figure 4.7 Evolution of the learning-based PI for each clock offset scenario.

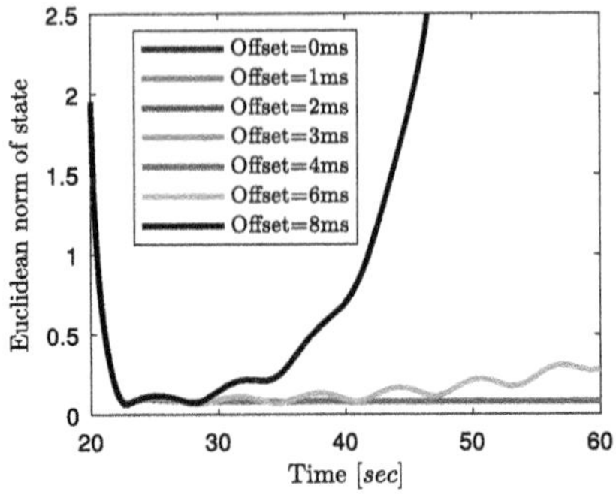

Figure 4.8 Evolution of $\|x(t)\|$ over $t \in [20,\ 60]$ for each offset scenario.

Proof. The proof follows an almost identical argument to that of Lemma 4.3 and Theorem 4.3. The only difference is that all limits are taken as $\rho \nearrow 1$ (in addition to $\Delta_M \searrow 0$ and $N_v, N_u, N_a \to \infty$) and that Lemma 4.4 is used to account for the uniform convergence of the quantization error to zero over Ω. □

4.3.8 Simulation results

We consider the following two-link manipulator of Modares et al. (2015):

$$M(q)\ddot{q} + V_m(q, \dot{q})\dot{q} + F_d\dot{q} + F_s(\dot{q}) = u + a,$$

where $q \in \mathbb{R}^2$ and $\dot{q} \in \mathbb{R}^2$ are the angular positions (in rad) and the angular velocities (in rad/s), respectively, and the state vector is defined as $x = [q^\mathrm{T}\ \dot{q}^\mathrm{T}]^\mathrm{T}$. The matrices $M(q) \in \mathbb{R}^{2\times2}$ and $V_m(q, \dot{q}) \in \mathbb{R}^{2\times2}$ are the inertia and the centripetal–Coriolis matrices, whereas $F_d\dot{q}$ and $F_s(\dot{q})$ model the dynamic and static frictions, respectively. The input $u \in \mathbb{R}^2$ denotes the torque, whereas $a \in \mathbb{R}^2$ is the adversarial input (both in Nm). The objective is to approximate the optimal game-based value function $V^\star$ and controller $u^\star$ of (4.35), where $Q(x) = 5\,\|x\|^2$, $R = I_2$, and $\gamma^2 = 20$. To this end, the actor-critic network (4.45) is employed with basis functions given by polynomials up to the fourth order.

First, we consider a setup where the input data sampled from the system suffer from clock offsets. To showcase the gradual deterioration in performance as the magnitude of these offsets increases, we simulate seven different cases, in each of which the clock offsets of the input signals $\delta_i^x(t)$ and $\delta_i^u(t)$ take the constant values 0 ms, 1 ms, 2 ms, 3 ms, 4 ms, 6 ms and 8 ms, respectively, for all $t \geq t_0$, $i = 1, 2$. In each simulation, the first 20 sec-

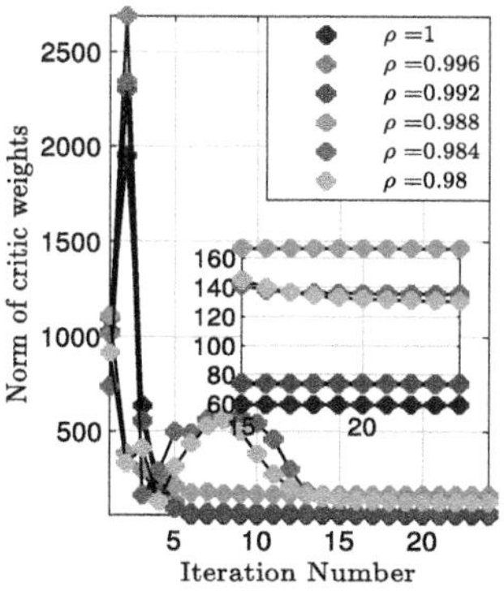

(a) Evolution of the norm of the critic weights at each iteration of the learning-based PI with quantization.

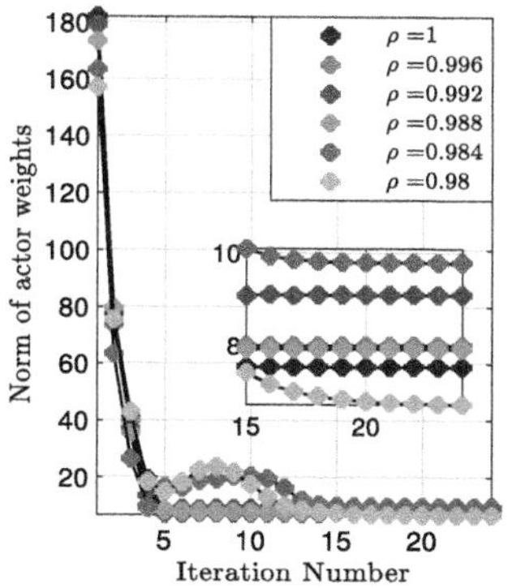

(b) Evolution of the norm of the actor weights at each iteration of the learning-based PI with quantization.

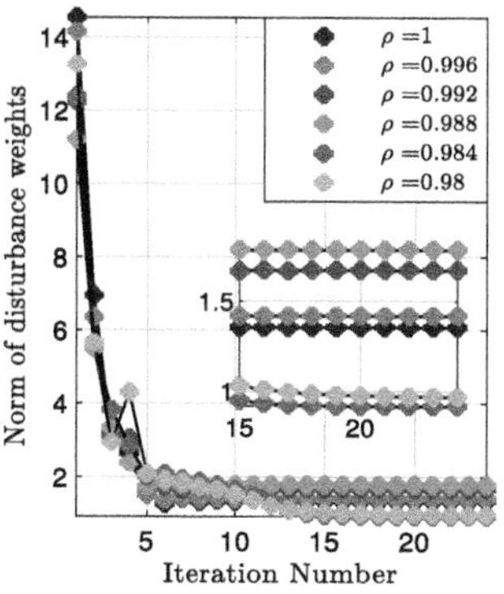

(c) Evolution of the norm of the disturbance weights at each iteration of the learning-based PI with quantization.

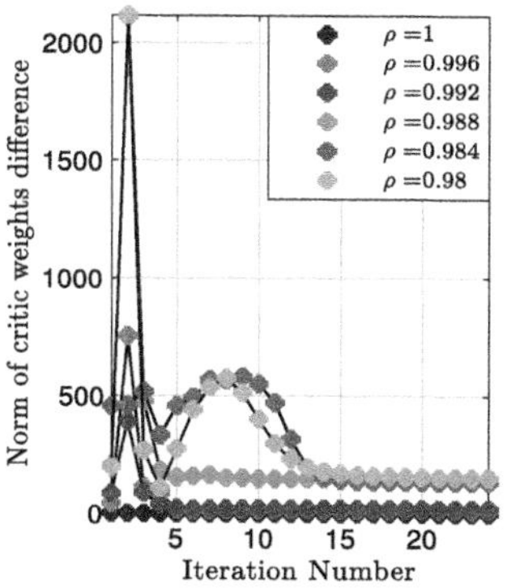

(d) Evolution of the norm of the difference of the critic weights at each scenario with the critic weights of the quantization-free scenario.

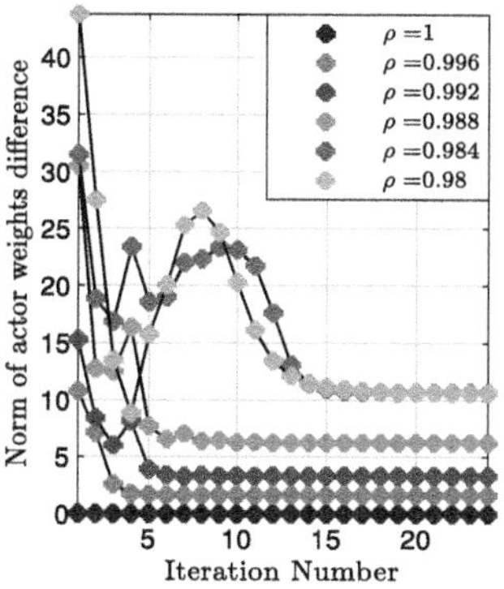

(e) Evolution of the norm of the difference of the actor weights at each scenario with the actor weights of the quantization-free scenario.

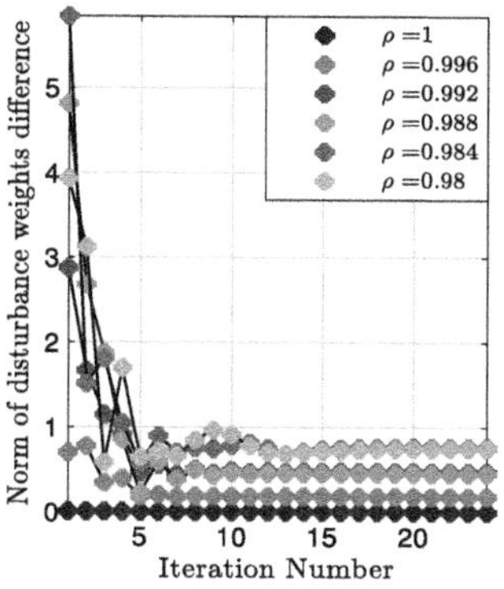

(f) Evolution of the norm of the difference of the disturbance weights at each scenario with the disturbance weights of the quantization-free scenario.

Figure 4.9 Evolution of the learning-based PI for each quantization scenario.

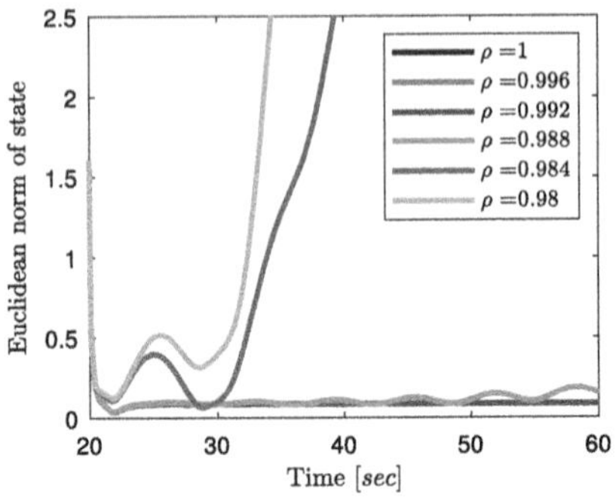

Figure 4.10 Evolution of $\|x(t)\|$ over $t \in [20,\ 60]$ for each quantization scenario.

onds are used for exploration to gather sufficient state-input data from the system. Subsequently, the learning-based PI with clock mismatches (Algorithm 4.6) is carried out with $T = 50\,\text{ms}$ by iteratively solving Eqs. (4.50). The learned controller is then implemented in the system for all $t \geq 20s$, where the adversarial input is $a(t) = 0.5\sin(t)[1\ \ 1]^{\mathrm{T}}$.

The results are shown in Figs. 4.7 and 4.8. We can see from Figs. 4.7(a)–4.7(c) that the convergence of the learning-based PI algorithm takes place for all values of the clock mismatch. However, as seen from Figs. 4.7(d)–4.7(e), the network weights converge monotonically further away from their nominal values (i.e., their values when the clock mismatch is zero) as the magnitude of the offset is increased. This behavior is expected by Theorem 4.3. Fig. 4.8 shows the evolution of the norm of the state vector for each clock mismatch scenario. Despite the persistent disturbance, the closed loop remains bounded for offsets ranging up to 4 ms. However, the system trajectories drift to infinity for offsets equal to 6–8 ms.

Second, we consider a setup where the state data sampled from the system suffer from logarithmic quantization with $a_0 = 1$. We specifically simulate six different cases, in each of which the characteristic value ρ of the quantizer is 1, 0.996, 0.992, 0.988, 0.984, and 0.980, respectively; the value $\rho = 1$ is used with a slight abuse of notation to describe the quantization-free case. Similarly to the previous setup, the first 20 seconds are used for exploration, and subsequently the learning-based PI with quantization (Algorithm 4.7) is carried out with $T = 50\,\text{ms}$ by iteratively solving Eqs. (4.69). The learned controller is then implemented in the system for all $t \geq 20s$, where the adversarial input is $a(t) = 0.5\sin(t)[1\ \ 1]^{\mathrm{T}}$. The results present a behavior similar to the clock offset scenario: Although the convergence of the learning takes place for all quantizers as shown in Figs. 4.9(a)–4.9(c), the network weights converge monotonically further away from their nominal values as the quantizer becomes more coarse (Figs. 4.9(d)–4.9(f)). In addition, for the more coarse quantizers where $\rho \in \{0.980,\ 0.984,\ 0.988\}$, the system trajectories drift to infinity under the learned control law (Fig. 4.10).

CHAPTER FIVE

Adversarial modeling

5.1. Bounded rationality in dynamical environments

5.1.1 Non-equilibrium non-zero-sum differential game

Initially, we formulate the problem in the general case for non-equilibrium non-zero-sum differential games with N players. We denote by $\mathcal{N}$ the set of all players and by $\mathcal{N}_{-i} = \mathcal{N} \setminus \{i\}$ the set containing all players except player i. The environment evolves according to the following dynamics:

$$\dot{x} = f(x) + \sum_{j \in \mathcal{N}} g_j(x) u_j, \ \ x(0) = x_0, \ \ t \geq 0,$$

where $x(t) \in \mathbb{R}^n$ is the state vector, which is available for feedback to all the players, $u_i \in \mathbb{R}^m$, $i \in \mathcal{N}$, are the decision vectors, $f(x) : \mathbb{R}^n \to \mathbb{R}^n$ is the drift dynamics, and $g_i(x) : \mathbb{R}^n \to \mathbb{R}^m$, $i \in \mathcal{N}$, are the input dynamics.

Each player aims to minimize a cost functional of the form

$$J_i(x(0); u_1, \ldots, u_N) = \int_0^\infty r_i(x(t), u_1, \ldots, u_N) \mathrm{d}\tau$$
$$:= \int_0^\infty \Big(Q_i(x) + u_i^{\mathrm{T}} R_{ii} u_i + \sum_{j \in \mathcal{N}_{-i}} u_j^{\mathrm{T}} R_{ij} u_j \Big) \mathrm{d}\tau, \ \ i \in \mathcal{N},$$

where $Q_i(x) : \mathbb{R}^n \to \mathbb{R}$, $i \in \mathcal{N}$, is the state cost, and $R_{ij} \in \mathbb{R}^{m \times m}$, $i, j \in \mathcal{N}$, are the control cost matrices.

The derivation of the Nash policies $u_i^\star$, $i \in \{1, \ldots, N\}$, corresponds to solving the coupled minimization problems

$$J_i(x(0); u_i^\star, u_{-i}^\star) = \min_{u_i} J_i(x(0); u_i, u_{-i}^\star).$$

Toward that, we need to evaluate the optimal value functions

$$V_i(x) = \min_{u_i} \int_t^\infty r_i(x, u_i, u_{-i}^\star) \mathrm{d}\tau, \qquad i \in \{1, \ldots, N\}.$$

Control and Game Theoretic Methods for Cyber-Physical Security
https://doi.org/10.1016/B978-0-44-315408-9.00011-7

Initially, we define the Hamiltonian of each player as

$$\begin{aligned} H_i(x, u_i, u_{-i}) =& r_i(x(t), u_1, \ldots, u_N) \\ &+ \nabla V_i^{\mathrm{T}}\Big(f(x) + \sum_{j\in N} g_j(x)u_j\Big). \end{aligned} \tag{5.1}$$

To compute the Nash policy for every player $i \in \{1, \ldots, N\}$, we apply the stationarity conditions to (5.1), i.e., $\frac{\partial H}{\partial u_i} = 0$, to get

$$u_i^\star(x) = -R_{ii}^{-1} g_i^{\mathrm{T}}(x) \frac{\partial V_i^\star}{\partial x}. \tag{5.2}$$

Substituting (5.2) for all players $i \in N$ into (5.1) yields the following coupled Hamilton–Jacobi system of N equations (Vamvoudakis and Lewis, 2011):

$$\begin{aligned} &Q_i(x) + \frac{1}{4}\sum_{j\in N} \nabla V_j^{\mathrm{T}} g_j(x) R_{jj}^{-\mathrm{T}} R_{ij} R_{jj} g_j^{\mathrm{T}}(x) \nabla V_j \\ &+(\nabla V_i)^{\mathrm{T}}\Big(f(x) - \frac{1}{2}\sum_{j\in N} g_j(x) R_{jj}^{-1} g_j^{\mathrm{T}}(x)\nabla V_j\Big) = 0. \end{aligned}$$

In this work, we follow the approach of "steps of reasoning," where each player belongs to a specific "cognitive type."

Due to computational and cognitive limitations, the emergence of Nash solutions in experimental games is rare. In this work, we are interested in introducing opponents with bounded rationality in differential games for CPS security. Initially, we will define strategies of "level-k" thinkers as different steps of strategic thinking where a distribution of beliefs of the rest of the players is considered known. Subsequently, by limiting our analysis to two-player zero-sum games, we will introduce a fitting mechanism that enables estimation of the levels of rationality in the environment.

Consider a player $i \in N$ belonging to a cognitive $k \in \mathbb{Z}^+$. This player can perform k steps of iterative strategic thinking while their beliefs about the behavior of the rest of the players are constructed based on a fixed, full-support distribution, e.g. – according to the work of Camerer (2003) – a Poisson distribution.

Level-0 policies

Bounded rationality approaches to non-equilibrium games that depend on iterative best responses assume that all the players have prior knowledge of the strategies of those that do not conduct strategic thinking, i.e., level-0 policies u_i^0, $i \in \mathcal{N}$. Previous work on non-equilibrium game theory (Strzalecki, 2014) proposes two different definitions of anchor policies, random policies and intuitive responses to the game.

Level-*k* policies

For a level-k player to compute their higher-order policies, they follow a sequence of k thinking steps, which will be modeled through a policy iteration scheme. The level-k player initializes their thinking process by assuming that every other player uses their level-0 policies. Subsequently, the player computes the best response to those policies, which constitutes their level-1 policy. At each subsequent level, the player computes the best response according to the expected behavior of the other players. For example, a level-3 player best responds to the probability that each of the other players follows a level $l < 3$ policy, where those probabilities are sampled from a predefined distribution.

The single-agent optimization that a level-k player performs corresponds to solving the following Hamilton–Jacobi–Bellman (HJB) equation with respect to a non-equilibrium value function $V^k(x) : \mathbb{R}^n \to \mathbb{R}$, conditioned by the belief probabilities of the agent of level-k, regarding the intelligence level of player $h \in \mathcal{N} \setminus \{i\}$, denoted as $b^k(h)$:

$$Q(x) + u_i^{k\mathrm{T}} R_{ii} u_i^k + \sum_{j \in \mathcal{N}_{-i}} \sum_{h=0}^{k-1} (b^k(h) u_j)^{h\mathrm{T}} R_{ij} b^k(h) u_j^h$$
$$+ \left(\nabla V_i^k\right)^{\mathrm{T}} \left(f(x) + g(x) u_i^k + \sum_{j \in \mathcal{N}_{-i}} \sum_{h=0}^{k-1} (g(x) b^k(h) u_j)\right) = 0,$$

which yields the level-k policy given by

$$u_i^k(x) = -R_{ii}^{-1} g_i^{\mathrm{T}} \nabla V_i^k(x) \;\; \forall x.$$

Results for the convergence and stability properties of the cognitive hierarchy approach to non-zero-sum nonlinear games have been studied in Kokolakis et al. (2020) based on approaches developed in the context of this thesis. Thus here we will present our findings on zero-sum games with dynamics described by linear ordinary differential equations.

Consider a system that is under attack by adversarial agents with different levels of rationality,

$$\dot{x}(t) = Ax(t) + Bu(t) + Kd(t), \;\; x(0) = x_0, \;\; t \geq 0, \tag{5.3}$$

where $x(t) \in \mathbb{R}^n$ is the state vector available for feedback to all the players, $u(t) \in \mathbb{R}^m$ is the defender's input, $d(t) \in \mathbb{R}^k$ is the active adversarial input, and $A \in \mathbb{R}^{n\times n}$, $B \in \mathbb{R}^{n\times m}$, and $K \in \mathbb{R}^{n\times k}$ are the drift, input, and adversarial matrices, respectively.

Under the assumption of perfect rationality, we seek to find the saddle-point solution of a zero-sum game between the defender and an adversary d, based on the following functional:

$$J(x; u, d) = \frac{1}{2}\int_0^\infty \left(x^{\mathrm{T}}Mx + u^{\mathrm{T}}Ru - \gamma^2\|d\|^2\right) \mathrm{d}t,$$

where the user-defined matrices satisfy $M \succeq 0$, $R \succ 0$, and $\gamma > \gamma^\star > 0$, where $\gamma^\star$ is the minimal γ that renders the integral finite.

Assumption 5.1. The pair (A, B) is controllable, and the pair $(A, \sqrt{M})$ is observable. ■

Assumption 5.2. $BR^{-1}B^{\mathrm{T}} - \frac{1}{\gamma^2}KK^{\mathrm{T}} \succeq 0$. ■

Assumption 5.3. The system interacts with several attackers sequentially by playing a zero-sum with the active attacker, whereas multiple adversaries do not interact with the system simultaneously. ■

Given that both players are playing a zero-sum game, we are interested in solving the following min-max optimization problem:

$$V^\star(x_0) = \min_u \max_d \int_0^\infty \left(x^{\mathrm{T}}Mx + u^{\mathrm{T}}Ru - \gamma^2\|d\|^2\right) \mathrm{d}\tau. \tag{5.4}$$

Definition 5.1. The pair of policies $(u^\star, d^\star)$ is said to be a saddle point if $J(u^\star, d) \le J(u^\star, d^\star) \le J(u, d^\star)$ for all u, d. ■

The Hamiltonian associated with (5.4) and (5.3) is

$$H(x, u, d, \frac{\partial V^\star}{\partial x}) = x^{\mathrm{T}}Mx + u^{\mathrm{T}}Ru - \gamma^2\|d\|^2 + \left(\frac{\partial V^\star}{\partial x}\right)^{\mathrm{T}}(Ax + Bu + Kd), \ \forall x, u, d. \tag{5.5}$$

Applying the stationarity conditions in (5.5), namely $\frac{\partial H}{\partial u} = 0$ and $\frac{\partial H}{\partial d} = 0$, yields the policies

$$u^\star(x) = -R^{-1}B^{\mathrm{T}}\frac{\partial V^\star}{\partial x}, \tag{5.6}$$

$$d^{\star}(x) = \frac{1}{\gamma^2} K^{\mathrm{T}} \frac{\partial V^{\star}}{\partial x} \tag{5.7}$$

for the defender and adversary, respectively.

Expressing the value function (5.4) as quadratic in the state x, i.e., $V^{\star}(x) = x^{\mathrm{T}} P x$, $P \succ 0$, substituting (5.6) and (5.7) into (5.5), and recalling that $H(x, u^{\star}, d^{\star}, \frac{\partial V^{\star}}{\partial x}) = 0$ yield the following game Riccati equation:

$$A^{\mathrm{T}} P + PA - PBR^{-1} B^{\mathrm{T}} P + \frac{1}{\gamma^2} PKK^{\mathrm{T}} P + M = 0. \tag{5.8}$$

Under Assumptions 5.1 and 5.2, the Riccati equation (5.8) has a unique solution, which is described by the saddle point $(u^{\star}, d^{\star})$ of the zero-sum game.

In realistic scenarios, the assumption of infinite rationality for all players is rarely satisfied. Thus we are interested in formulating models for the attacking agents with bounded rationality in the context of zero-sum differential games and introducing such framework to identify the true distribution of intelligence levels in an adversarial environment. Initially, we will define the strategies of "level-k" thinkers as different steps of strategic thinking. Subsequently, we will allow the true attack input to be arbitrary and show how we can fit the measured attacks to the different levels of rationality that have been computed.

5.1.2 Levels of rationality

We present an iterative method that will describe the policies of the players performing k steps of strategic thinking.

Level-0 (anchor) policy

To define different levels of rationality, we need to introduce an anchor policy for the level-0 player. In the literature of non-equilibrium games (Strzalecki, 2014), there are two approaches for defining level-0. The policy is either considered to be a random one or an intuitive response to the game. To avoid the stochastic effects of a random input, we follow the latter. We will define the level-0 defender strategy as the policy in the absence of attacks. Consequently, the level-0 input is found by solving an optimal control problem described by

$$V_u^0(x_0) = \min_u \int_0^{\infty} (x^{\mathrm{T}} Mx + u^{\mathrm{T}} Ru) \mathrm{d}\tau. \tag{5.9}$$

The optimal control input for the optimization problem (5.9) given (5.3) with $d = 0$ is

$$u^0(x) = -R^{-1}B^{\mathrm{T}}\frac{\partial V_u^0(x)}{\partial x} = -R^{-1}B^{\mathrm{T}}P_u^0 x \ \forall x,$$

where the value function is taken as quadratic in the states $V_u^0(x) = x^{\mathrm{T}}P_u^0 x$, and the kernel P_u^0 solves the Riccati equation $A^{\mathrm{T}}P_u^0 + P_u^0 A + M - P_u^0 BR^{-1}B^{\mathrm{T}}P_u^0 = 0$.

Subsequently, the intuitive response of a level-0 adversary is an optimal attack under the belief that the defender is unaware of their existence. To this end, we define the optimization problem from the point of view of the adversary for the anchor defender input $u = u^0(x)$,

$$V_d^0(x_0) = \max_d \int_0^\infty (x^{\mathrm{T}}Mx + u^{0\mathrm{T}}Ru^0 - \gamma^2\|d\|^2)\mathrm{d}\tau,$$

subject to $\dot{x} = (A - BR^{-1}B^{\mathrm{T}}P_u^0)x + Kd$.

The level-0 attack input is computed as $d^0(x) = \frac{1}{\gamma^2}K^{\mathrm{T}}P_d^0 x$, where the matrix P_d^0 solves the Riccati equation

$$\begin{aligned} 0 = &(A - BR^{-1}B^{\mathrm{T}}P_u^0)^{\mathrm{T}}P_d^0 + P_d^0(A - BR^{-1}B^{\mathrm{T}}P_u^0) \\ &+ (M + P_u^0 BR^{-1}B^{\mathrm{T}}P_u^0) + \frac{1}{\gamma^2}P_d^0 KK^{\mathrm{T}}P_d^0. \end{aligned}$$

Level-k policies

To derive the policies for the agents of higher levels of rationality, we will follow an iterative procedure, wherein the defender and adversary optimize their respective strategies under the belief that their opponent is using a lower level of thinking. A defender performing an arbitrary number of k strategic thinking interactions will derive the optimal policy $u^k(x)$ according to the respective k-level value function $V_u^k(x)$. Similarly, the k-level policy of the attacker is given as $d^k(x)$, alongside their respective k-level value function $V_d^k(x)$.

The defender's kth policy is given as the solution to the following minimization problem:

$$V_u^k(x_0) = \min_u \int_0^\infty (x^{\mathrm{T}}Mx + u^{\mathrm{T}}Ru - \gamma^2\|d^{k-1}\|^2)\mathrm{d}\tau \tag{5.10}$$

subject to the constraint

$$\dot{x} = Ax + Bu + Kd^{k-1}.$$

The corresponding Hamiltonian is

$$H_u^k(x, u, d^{k-1}) = x^{\mathrm{T}}Mx + u^{\mathrm{T}}Ru - \gamma^2\|d^{k-1}\|^2 + \left(\frac{\partial V_u^k}{\partial x}\right)^{\mathrm{T}}\left(Ax + Bu + Kd^{k-1}\right) \ \forall x, u. \tag{5.11}$$

Substituting the adversarial input with the policy of the previous level $d^{k-1} = \frac{1}{\gamma^2}K^{\mathrm{T}}P_d^{k-1}x$, starting from level-0, yields

$$u^k(x) = -R^{-1}B^{\mathrm{T}}P_u^k x \ \forall x, \tag{5.12}$$

where the level-k defender Riccati matrix is the solution of

$$0 = (A + \frac{1}{\gamma^2}KK^{\mathrm{T}}P_d^{k-1})^{\mathrm{T}}P_u^k + P_u^k(A + \frac{1}{\gamma^2}KK^{\mathrm{T}}P_d^{k-1}) + (M - \frac{1}{\gamma^2}P_d^{k-1}KK^{\mathrm{T}}P_d^{k-1}) - P_u^kBR^{-1}B^{\mathrm{T}}P_u^k. \tag{5.13}$$

Similarly, the adversary of an arbitrary k level of thinking maximizes their response to the input of a defender of level-k:

$$V_d^k(x_0) = \max_d \int_0^\infty (x^{\mathrm{T}}Mx + \left(u^k\right)^{\mathrm{T}}Ru^k - \gamma^2\|d\|^2)\mathrm{d}\tau \tag{5.14}$$

subject to

$$\dot{x} = Ax + Bu^k + Kd.$$

The corresponding Hamiltonian is

$$H_d^k(x, u^k, d) = x^{\mathrm{T}}Mx + \left(u^k\right)^{\mathrm{T}}Ru^k - \gamma^2\|d\|^2 + \left(\frac{\partial V_d^k}{\partial x}\right)^{\mathrm{T}}\left(Ax + Bu^k + Kd\right) \ \forall x, d. \tag{5.15}$$

Substituting (5.12) into (5.15) yields the following attack response:

$$d^k(x) = \frac{1}{\gamma^2}K^{\mathrm{T}}P_d^k x \ \forall x, \tag{5.16}$$

where the matrix P_d^k is the solution of

$$0 = (A - BR^{-1}B^{\mathrm{T}}P_u^k)^{\mathrm{T}}P_d^k + P_d^k(A - BR^{-1}B^{\mathrm{T}}P_u^k) + (M + P_u^kBR^{-1}B^{\mathrm{T}}P_u^k) + \frac{1}{\gamma^2}P_d^kKK^{\mathrm{T}}P_d^k. \tag{5.17}$$

With this iterative procedure, a defending agent can compute the strategies of the adversaries with finite cognitive abilities for a given number of levels.

Remark 5.1. Since $R \succ 0$, the elements of the Hessian matrix corresponding to the defender for the optimization problem solved by (5.13) are positive definite, and therefore the defender minimizes. Similarly, since $\gamma > 0$, (5.17) defines a maximization problem. ■

Theorem 5.1. *Consider the pairs of strategies at a specific cognitive level-k, given by* (5.12) *and* (5.13) *for the defender and by* (5.16) *and* (5.17) *for the adversary. The policies converge to a Nash equilibrium for higher levels if the following conditions hold as the levels increase:*

$$(P_d^{k-1} + P_d^k - P_u^{k+1})KK^T(P_d^k - P_d^{k-1}) \succ 0, \tag{5.18}$$

$$(3P_u^{k-1} - P_u^k + P_d^k)BR^{-1}B^T(P_u^k - P_u^{k-1}) \succ 0. \tag{5.19}$$

Proof. We will consider the problem from the point of view of the defender. We notice that at the $(k+1)$th thinking step, the best response of the player corresponds to the Hamiltonian

$$H_u^{k+1}(x, u^{k+1}, d^k) = x^T Mx + \left(u^{k+1}\right)^T Ru^{k+1} - \gamma^2 \|d^k\|^2 + \left(\frac{\partial V_u^{k+1}}{\partial x}\right)^T \left(Ax + Bu^{k+1} + Kd^k\right) = 0 \;\; \forall x.$$

Thus, introducing the adversarial input d^{k-1}, we get

$$H_u^{k+1}(x, u^{k+1}, d^k) = x^T Mx + \left(u^{k+1}\right)^T Ru^{k+1} - \gamma^2 \|d^k + d^{k-1} - d^{k-1}\|^2 + \left(\frac{\partial V_u^{k+1}}{\partial x}\right)^T (Ax + Bu^{k+1} + K(d^k + d^{k-1} - d^{k-1})) \Rightarrow$$

$$\left(\frac{\partial V_u^{k+1}}{\partial x}\right)^T (Ax + Bu^{k+1} + Kd^{k-1}) + x^T Mx + \left(u^{k+1}\right)^T Ru^{k+1} - \gamma^2 \|d^{k-1}\|^2 - 2\gamma^2 (d^{k-1})^T (d^k - d^{k-1}) - \gamma^2 \|d^k - d^{k-1}\|^2 + \left(\frac{\partial V_u^{k+1}}{\partial x}\right)^T K(d^k - d^{k-1}) = 0.$$

Therefore the time derivative of the value function of the $(k+1)$th level defender V_u^{k+1} is

$$\dot{V}_u^{k+1} = -x^T Mx - \left(u^{k+1}\right)^T Ru^{k+1} + \gamma^2 \|d^{k-1}\|^2 - 2\gamma^2 (d^{k-1})^T (d^k - d^{k-1})$$

$$-\gamma^2\|d^k - d^{k-1}\|^2 + \left(\frac{\partial V_u^{k+1}}{\partial x}\right)^{\mathrm{T}} K(d^k - d^{k-1}) \ \ \forall x.$$

Thus, since it is known that

$$\dot{V}_u^k = -x^{\mathrm{T}}Mx - \left(u^k\right)^{\mathrm{T}} Ru^k + \gamma^2\|d^{k-1}\|^2 \ \ \forall x,$$

it follows that

$$\begin{aligned}\dot{V}_u^k \leq \dot{V}_u^{k+1} &+ 2\gamma^2(d^{k-1})^{\mathrm{T}}(d^k - d^{k-1}) + \gamma^2\|d^k - d^{k-1}\|^2 \\ &- \left(\frac{\partial V_u^{k+1}}{\partial x}\right)^{\mathrm{T}} K(d^k - d^{k-1}).\end{aligned}$$

To ensure that $\dot{V}_u^k \leq \dot{V}_u^{k+1}$, we need to satisfy

$$2\gamma^2(d^{k-1})^{\mathrm{T}}(d^k - d^{k-1}) + \gamma^2\|d^k - d^{k-1}\|^2 - \left(\frac{\partial V_u^{k+1}}{\partial x}\right)^{\mathrm{T}} K(d^k - d^{k-1}) > 0. \tag{5.20}$$

Furthermore, substituting (5.16) into (5.20) and simplifying the derived expression yield (5.18). Thus by integrating the two sides of the inequality we get

$$V_u^{k+1} \leq V_u^k.$$

By the principle of optimality the value functions $V_u^k(x_0)$ are lower bounded by (5.4). The iterations of $V_u^k(x_0)$ converge to the Nash value function, and, by extension, the defender's strategy converges to the Nash policy. Identical arguments can be made for the adversary, where we require (5.19) to be satisfied. □

Remark 5.2. Note that the proposed framework for non-equilibrium behavior in differential games leverages results from reinforcement learning. The performance of thinking steps as presented in level-*k* and cognitive hierarchy models has clear connections to policy iteration algorithms. Thus the proof of Theorem 5.1 is inspired by and closely follows the reinforcement learning literature (Beard and McLain, 1998; Vamvoudakis et al., 2012). ■

The integral cost functionals need to be finite for a solution to exist. As a result, the adversary must remain stealthy and unable to destabilize the system, a condition which is captured by the following theorem.

Theorem 5.2. *Consider system (5.3) under the effect of agents with bounded rationality whose policies are defined by (5.12) for the defender and by (5.16) for the adversary. The game can be solved up to any cognitive level-k as long as*

$$P_u^k B R^{-1} B^T P_u^k \succ \max\left(\frac{1}{\gamma^2} P_d^{k-1} K K^T P_d^{k-1}, \frac{1}{\gamma^2} P_d^k K K^T P_d^k\right), \qquad (5.21)$$

where $\max(\cdot,\cdot)$ *is an operator that maps to its argument with the largest value.*

Proof. Initially, we will consider the optimal control of an arbitrary level-k from the perspective of the defender, as defined by (5.10). For the system to have a solution, the value function must be a Lyapunov function whose orbital derivative, according to (5.11), satisfies

$$\left(\frac{\partial V_u^k}{\partial x}\right)^{\mathrm{T}} \left(Ax + Bu + Kd^{k-1}\right) = -(x^{\mathrm{T}} M x + u^{\mathrm{T}} R u - \gamma^2 \|d^{k-1}\|^2) \ \forall x, u.$$

Substituting the policies (5.12) and (5.16) yields

$$\left(\frac{\partial V_u^k}{\partial x}\right)^{\mathrm{T}} \left(Ax + Bu + Kd^{k-1}\right) = -x^{\mathrm{T}} M x - x^{\mathrm{T}} (P_u^k B R^{-1} B^{\mathrm{T}} P_u^k - \frac{1}{\gamma^2} P_d^{k-1} K K^T P_d^{k-1}) x \ \forall x.$$

To guarantee the asymptotic stability of the equilibrium point of the closed-loop system and, by extension, a finite value for the cost integral, the following condition needs to hold:

$$P_u^k B R^{-1} B^{\mathrm{T}} P_u^k \succ \frac{1}{\gamma^2} P_d^{k-1} K K^T P_d^{k-1}.$$

Applying the same analysis to the optimal control problem from the perspective of the adversary with the same level of intelligence (see (5.14)), we require that

$$P_u^k B R^{-1} B^{\mathrm{T}} P_u^k \succ \frac{1}{\gamma^2} P_d^k K K^{\mathrm{T}} P_d^k,$$

completing the proof. □

Remark 5.3. Theorem 5.2 states that since the opponents take turns in calculating their best responses, the defender's policy u^k at a certain level-k must be able to stabilize the system under the attack input to which it responds, i.e., d^{k-1}, but also under the worst-case response to u^k, i.e., d^k. This will hold if γ is large enough to satisfy (5.21) for every level k. ■

5.1.3 Model-free learning for interaction with level-k attacks

By leveraging our previous work on Q-learning (Vamvoudakis, 2017) we can learn the best responses of all level-k agents without explicit knowledge of the physics of the system. Model-free learning requires the development of an action-dependent value function that has the same solution as the state-dependent value function (Sutton and Barto, 2018). We can define this action-dependent function in various ways, through the use of integral reinforcement learning (IRL) (Luo et al., 2014; Vrabie et al., 2013) or by constructing an appropriate action-dependent function that is learned through IRL (Vamvoudakis, 2017).

Consider the problem solved by an agent, described by (5.10) for the defender and (5.14) for the adversary. For ease of exposition, we will introduce the notations $V_j^k(\cdot)$, $a_j^k(x)$, and $H_j^k(x, a_j^k, \nabla V_j^k)$ to mean the value function, action policy, and Hamiltonian of a level-k agent, where $j \in \{u, d\}$, i.e., the defender and the adversary.

The action-dependent function can be defined as

$$Q_j^k(x, a_j^k) := V_j^k(x) + H_j^k(x, a_j^k, \nabla V_j^k) \;\; \forall x, a_j^k,\; j \in \{u, d\}. \tag{5.22}$$

Thus (5.22) is an advantage function whose the most important property is that it shares the same minimum as the value function (Baird, 1994; Bradtke et al., 1994). Substituting the Hamiltonians (5.11) for the defender and (5.15) for the adversary into (5.22), we can rewrite the action-dependent function (5.22) in a compact quadratic in the state and action form as

$$Q_j^k(x, a_j^k) = \left(U_j^k\right)^{\mathrm{T}} \begin{bmatrix} Q_{j,\mathrm{xx}}^k & Q_{j,\mathrm{xa}}^k \\ Q_{j,\mathrm{ax}}^k & Q_{j,\mathrm{aa}}^k, \end{bmatrix} U_j^k := \left(U_j^k\right)^{\mathrm{T}} \tilde{Q}_j^k U_j^k \;\; \forall x, a_j^k,\; j \in \{u, d\}, \tag{5.23}$$

where for the level-k defender's problem, $j := u$, we have that $U_u^k = \begin{bmatrix} x^{\mathrm{T}} & (u^k)^{\mathrm{T}} \end{bmatrix}^{\mathrm{T}}$,

$$\begin{aligned} Q_{u,\mathrm{xx}}^k &= (A + \frac{1}{\gamma^2} K K^{\mathrm{T}} P_d^{k-1})^{\mathrm{T}} P_u^k + P_u^k (A + \frac{1}{\gamma^2} K K^{\mathrm{T}} P_d^{k-1}) \\ &\quad + (M - \frac{1}{\gamma^2} P_d^{k-1} K K^{\mathrm{T}} P_d^{k-1}) - P_u^k B R^{-1} B^{\mathrm{T}} P_u^k + P_u^k, \\ Q_{u,\mathrm{xa}}^k &= B^{\mathrm{T}} P_u^k,\;\; Q_{u,\mathrm{ax}}^k = B P_u^k,\;\; Q_{u,\mathrm{aa}}^k = R. \end{aligned}$$

Accordingly, for the level-k adversarial problem, $j := d$, we have that $U_d^k = [x^{\mathrm{T}} \quad (d^k)^{\mathrm{T}}]^{\mathrm{T}} z$,

$$\begin{aligned} Q_{d,\mathrm{xx}}^k &= (A - BR^{-1}B^{\mathrm{T}}P_u^k)^{\mathrm{T}}P_d^k + P_d^k(A - BR^{-1}B^{\mathrm{T}}P_u^k) \\ &\quad + (M + P_u^k BR^{-1}B^{\mathrm{T}}P_u^k) + \frac{1}{\gamma^2}P_d^k KK^{\mathrm{T}}P_d^k, \\ Q_{d,\mathrm{xa}}^k &= K^{\mathrm{T}}P_d^k, \; Q_{d,\mathrm{ax}}^k = KP_d^k, \; Q_{d,\mathrm{aa}}^k = -\gamma^2. \end{aligned}$$

The action for each agent at every level can be found by solving $\frac{\partial Q_j^k(x, a_j^k)}{\partial a_j^k} = 0$ as

$$a_j^k(x) = -(Q_{j,\mathrm{aa}}^k)^{-1} Q_{j,\mathrm{ax}}^k x \;\; \forall x, \; j \in \{u, d\}. \tag{5.24}$$

We will now use an actor/critic structure to tune the parameters online by utilizing an integral reinforcement learning approach (Vrabie et al., 2013). The level-k critic approximator will approximate the action-dependent function (5.23), whereas the level-k actor will approximate the appropriate defense or attack policy (5.24).

To do that, we write the action-dependent function as

$$Q_j^k(x, a_j^k) = \mathrm{vech}(\tilde{Q})^{\mathrm{T}}(U_j^k \otimes U_j^k) \;\; \forall x, a_j^k, \; j \in \{u, d\}, \tag{5.25}$$

where the Kronecker product-based polynomial quadratic polynomial function $(U_j^k \otimes U_j^k)$ is reduced to guarantee linear independence of the elements. The vectorized action-dependent function (5.25) can be described in terms of the ideal weights $W_j^k = \mathrm{vech}(\tilde{Q})$, leading to the compact form $Q_j^k(x, a_j^k) = (W_j^k)^{\mathrm{T}}(U_j^k \otimes U_j^k)$.

Since the ideal weights are unknown, we will consider the following estimated level-k action-dependent function according to the estimated critic weights $\hat{W}_j^k = \mathrm{vech}(\hat{\tilde{Q}})$:

$$\hat{Q}_j^k(x, a_j^k) = \left(\hat{W}_j^k\right)^{\mathrm{T}}(U_j^k \otimes U_j^k) \;\; \forall x, a_j^k, \; j \in \{u, d\}, \tag{5.26}$$

as well as the estimated actor approximator

$$\hat{a}_j^k(x) = \left(\hat{W}_{a,j}^k\right)^{\mathrm{T}} x \;\; \forall x, \; j \in \{u, d\}, \tag{5.27}$$

where the state x serves as the basis for the actor approximator, and $\hat{W}_{a,j}^k$ denotes the weight estimate of the level-k agent's policy.

The action-dependent function with the ideal weights, (5.23), has been shown (Vamvoudakis, 2017) to satisfy the integral form of the Hamilton–Jacobi–Bellman equation

$$Q_j^k(x(t), a_j^k(t)) = Q_j^k(x(t - T_{\mathrm{IRL}}), a_j^k(t - T_{\mathrm{IRL}})) - \int_{t-T_{\mathrm{IRL}}}^{t} \left(x^{\mathrm{T}} \bar{M}_j^k x + (a_j^k)^{\mathrm{T}} \bar{R}_j a_j^k\right) \mathrm{d}\tau, \ j \in \{u, d\}, \ t \geq 0,$$

where $T_{\mathrm{IRL}} \in \mathbb{R}^+$ is the sampling period of the algorithm. For the defender, we define $\bar{M}_u^k := M - \frac{1}{\gamma^2} P_d^{k-1} K K^{\mathrm{T}} P_d^{k-1}$ and $\bar{R}_u := R$. Similarly, for the adversary, we define $\bar{M}_d^k := M + P_u^k B R^{-1} B^{\mathrm{T}} P_u^k$ and $\bar{R}_d := -\gamma^2$.

We will now define the error based on the current estimate of the action-dependent function that we wish to drive to zero as

$$\begin{aligned} e_j^k &= \hat{Q}_j^k(x(t), a_j^k(t)) - \hat{Q}_j^k(x(t - T_{\mathrm{IRL}}), a_j^k(t - T_{\mathrm{IRL}})) \\ &\quad + \int_{t-T_{\mathrm{IRL}}}^{t} \left(x^{\mathrm{T}} \bar{M}_j^k x + \left(a_j^k\right)^{\mathrm{T}} \bar{R}_j a_j^k\right) \mathrm{d}\tau \\ &= \left(\hat{W}_j^k\right)^{\mathrm{T}} (U_j^k(t) \otimes U_j^k(t)) - \left(\hat{W}_k^k\right)^{\mathrm{T}} (U_j^k(t - T_{\mathrm{IRL}}) \otimes U_j^k(t - T_{\mathrm{IRL}})) \\ &\quad + \int_{t-T_{\mathrm{IRL}}}^{t} \left(x^{\mathrm{T}} \bar{M}_j^k x + \left(a_j^k\right)^{\mathrm{T}} \bar{R}_j a_j^k\right) \mathrm{d}\tau, \end{aligned}$$

as well as the policy error $e_{j,a}^k = (W_{a,j}^k)^{\mathrm{T}} x + (\hat{Q}_{j,\mathrm{aa}}^k)^{-1} \hat{Q}_{j,\mathrm{ax}} x$ for all x, $j \in \{u, d\}$, where the appropriate elements of the $\hat{Q}$ matrix will be extracted from the critic estimate $\hat{W}_j^k$.

Defining the squared error functions for the critic and actor weights $K_1 = \frac{1}{2}\|e_j^k\|^2$ and $K_2 = \frac{1}{2}\|e_{j,a}^k\|^2$ respectively, we derive the tuning rules by applying normalized gradient descent as

$$\dot{\hat{W}}_j^k = -\alpha \frac{\sigma_j^k}{\left(1 + (\sigma_j^k)^{\mathrm{T}} \sigma_j^k\right)^2} (e_j^k)^{\mathrm{T}}, \ t \geq 0, \ j \in \{u, d\}, \tag{5.28}$$

$$\dot{\hat{W}}_{j,a}^k = -\alpha_a x (e_{j,a}^k)^{\mathrm{T}}, \ t \geq 0, \ j \in \{u, d\}, \tag{5.29}$$

where $\sigma_j^k = (U_j^k(t) \otimes U_j^k(t)) - (U_j^k(t - T_{\mathrm{IRL}}) \otimes U_j^k(t - T_{\mathrm{IRL}}))$, and α, $\alpha_a \in \mathbb{R}^+$ are tuning gains.

Lemma 5.1. *Consider the costs for the defender and attacker,* (5.10) *and* (5.14), *respectively. The training of a level-k agent is described by the action-dependent function approximator* (5.26) *and their optimal policy by* (5.27). *The tuning rules for the critic and actor are given by* (5.28) *and* (5.29), *respectively. Then the closed-loop system of the augmented state* $\psi = \begin{bmatrix} x^T & (\hat{W}_j^k - W_j^k)^T & (\hat{W}_{j,a}^k - W_{j,a}^k)^T \end{bmatrix}^T$ *is asymptotically stable at the origin if the critic gain is picked sufficiently larger than the actor gain,*

$$1 < \alpha_a < \frac{1}{\delta \bar{\lambda}(\bar{R}^{-1})}(2\underline{\lambda}(\bar{M}_j^k + Q_{j,xa}^k \bar{R}^{-1}(Q_{j,xa}^k)^T) - \bar{\lambda}(Q_{j,xa}^k(Q_{j,xa}^k)^T)),$$

where $\delta \in \mathbb{R}^+$, *provided that for a given exploration period* $T_{exp} \in \mathbb{R}^+$, *the signal* $\Delta = \frac{\sigma_j^k}{(1+(\sigma_j^k)^T \sigma_j^k)^2}$ *is persistently excited over a period* $[t, t + T_{exp}]$, *i.e.,* $\int_t^{t+T_{exp}} \Delta \Delta^T d\tau \geq \beta I$ *with* $\beta \in \mathbb{R}^+$ *and the identity matrix* I *of appropriate dimension.*

Proof. The proof follows from Vamvoudakis (2017). □

Remark 5.4. The persistence of excitation condition can be guaranteed by adding exploration noise to the control input (Ioannou and Sun, 1996). ■

Remark 5.5. Although the described algorithm does not require explicit knowledge of the system matrices, both opponents require knowledge of the policies of the previous levels. This result agrees with similar claims from non-equilibrium game theory about the difficulty of operating in higher cognitive levels since for an agent to learn a level-*k* best response, first, she must successfully learn all the previous best responses of all the other agents acting on the system. ■

5.1.4 Estimation of the adversarial levels

In this section, we will propose an algorithmic framework where a defender interacts with adversaries of different cognitive levels for a predefined time period $T_{\text{int}} \in \mathbb{R}^+$. To further generalize our results, we will not restrict the adversaries to statically use only a specific level policy throughout the interaction. This will account for opponents who manage to increase their cognitive level after their initial attack, who try to deceive the defender or even the adversaries that do not follow a best response policy. Also, as we noted in Subsection 5.1.3, the defender computes the level-*k* responses from a model-free learning algorithm utilizing fictitious inputs.

Consequently, instead of identifying the exact distribution of the levels of intelligence in the adversarial environment, we fit an arbitrary attack input to the set of beliefs over the different level-k model-free policies.

Initially, we assume that the defender can directly measure the effect of the attack input on the system. We define the error between the actual measured adversarial input, denoted $f(t)$, and the adversarial input of a level-k adversary as

$$r^k = \int_{t-T_{\text{int}}}^{t} \|f(\tau) + (Q^k_{d,\text{aa}})^{-1} Q_{d,\text{xa}} x(\tau)\| \mathrm{d}\tau, \ t \geq 0, \ k \in \{1, \ldots, \mathcal{K}\}, \tag{5.30}$$

where $\mathcal{K}$ is the maximum level the defender has computed. We note that (5.30) is the norm of the "distance" of the measured attack from each cognitive level during the time of interaction with the particular adversary.

Corollary 5.1. *Consider an adversary operating at the Nash equilibrium, i.e.,*

$$f(t) = \frac{1}{\gamma^2} K^T P^\star x, \ t \geq 0. \tag{5.31}$$

Suppose that the hypotheses and statements of Lemma 5.1 *and Theorem* 5.1 *hold. Then* $\lim_{k\to\infty} r^k = 0$.

Proof. According to Lemma 5.1, any arbitrary level-k can be trained to converge to the best response strategy of an adversary (5.16), i.e., $(Q^k_{d,\text{aa}})^{-1} Q_{d,\text{xa}} x = \frac{1}{\gamma^2} K^{\mathrm{T}} P^k_d x$. Furthermore, if Theorem 5.1 holds, then as the cognitive level-k goes to infinity, it converges to the Nash solution, i.e., $\lim_{k\to\infty} \|P^k_d - P^\star\| = 0$. Consequently,

$$\lim_{k\to\infty} (-(Q^k_{d,\text{aa}})^{-1} Q_{d,\text{xa}} x) = \frac{1}{\gamma^2} K^{\mathrm{T}} P^\star x. \tag{5.32}$$

Substituting (5.31) and (5.32) into (5.30) as $k \to \infty$ proves the required result. □

After each interaction period $[t - T_{\text{int}}, t]$, we stack the elements r^k into the vector $\mathbf{r} = \begin{bmatrix} r^1 & r^2 & \ldots & r^{\mathcal{K}} \end{bmatrix}$. Motivated by Sutton and Barto (1998), we will use the softmax function to map the error vector $\mathbf{r}$ to a reinforcement-like signal. However, we need to reward those elements of the vector that are closer to the appropriate level, i.e., to reinforce the minimum element. As a result, we apply the softmax function to each element of $\mathbf{r}$ as follows:

$$\sigma^k = \frac{e^{-r^k}/\tau}{\sum_{i=1}^{\mathcal{K}} e^{-r^i}/\tau}, \tag{5.33}$$

where $\tau \in \mathbb{R}^+$ is the softmax temperature parameter. This will give us $\boldsymbol{\sigma} = [\sigma^1\ \sigma^2\ \dots\ \sigma^{\mathcal{K}}]^{\mathrm{T}}$. We model the distribution of the players over the different levels as a Poisson distribution with the probability mass function $p^k = \frac{\lambda^k e^{-\lambda}}{k!}$, $\lambda \in \mathbb{R}$. This probability will also define the belief of the defender about the relative proportion of level-k adversaries as

$$b^k = \frac{p^k}{\sum_{i=1}^{\mathcal{K}} p^i}. \tag{5.34}$$

Our goal is to evaluate the parameter λ. To this end, we use the following update rule for the mean of the observations:

$$\lambda^+ = \lambda + \frac{(\tilde{\mathbf{K}}^{\mathrm{T}}\boldsymbol{\sigma})n}{n+1}, \tag{5.35}$$

where $\tilde{\mathbf{K}} = [1\ 2\ \dots\ \mathcal{K}]$ is a vector containing the indexes of the levels we have trained, and n is the number of different agents we have interacted with.

5.1.5 Algorithmic framework for level distribution learning

The sequential interactions between the adversaries and defender as well as the estimation of the intelligence level distribution are shown in Fig. 5.1. The framework to learn the distribution of the different levels of the adversaries is presented in Algorithm 5.1.

Remark 5.6. The use of the softmax function and the "weighted observation" will allow for attack functions that do not follow the exact level-k policies we have trained the defender to expect. A more conservative approach would be to simply choose the minimum distance r^k and classify the attacker as one of the corresponding level-k thinkers. By calculating the mean level $\tilde{\mathbf{K}}^{\mathrm{T}}\mathbf{r}$ we can create a fictitious belief of the level of the adversaries that would capture their behavior even if they do not follow the calculated best response policies. By choosing a small temperature parameter τ we are able to classify the adversary to the level that corresponds to the minimum distance r^k. ■

Algorithm 5.1 Intelligence Level Learning.

1: **procedure**

2: Given initial state x_0, cost weights M, R, γ, highest allowable level defined $\mathcal{K}$, and time window T_{IRL}.

3: **for** $k = 0, \ldots, \mathcal{K}$ **do**

4: Set $j := u$ to learn the level-k defender policy.

5: Start with an initial guesses for $\hat{W}_u^k$, $\hat{W}_{u,a}^k$.

6: Propagate the augmented system with states $\chi = \left[x^{\mathrm{T}} \ \left(\hat{W}_u^k\right)^{\mathrm{T}} \ \left(\hat{W}_{u,a}^k\right)^{\mathrm{T}}\right]^{\mathrm{T}}$, according to (5.3), (5.28), and (5.29) until convergence.

7: Set $j := d$ to learn the level-k adversarial policy.

8: Start with initial guesses for $\hat{W}_d^k$, $\hat{W}_{d,a}^k$.

9: Propagate the augmented system with states $\chi = \left[x^{\mathrm{T}} \ \left(\hat{W}_d^k\right)^{\mathrm{T}} \ \left(\hat{W}_{d,a}^k\right)^{\mathrm{T}}\right]^{\mathrm{T}}$ according to (5.3), (5.28), and (5.29) until convergence. Go to 3.

10: **end for**

11: Define the interaction time with each adversary as T_{int}, the number of total interactions n_{int}, and an initial guess for λ.

12: **for** $i = 1, \ldots, n_{\mathrm{int}}$ **do**

13: For $t \in [t_i - T_{\mathrm{int}}, t_i]$, measure the value of (5.30).

14: Compute the mean level according to (5.33).

15: Update λ based on (5.35). Go to 13 to interact with a different adversary.

16: **end for**

17: **end procedure**

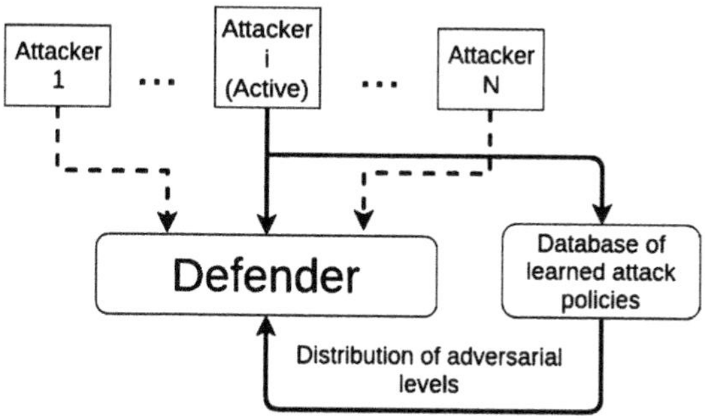

Figure 5.1 At each time instant, the active attacker's action is compared to the learned policies of the different levels. Once the defender has had an adequate input from the attackers, she can derive the distribution of the adversarial levels.

5.1.6 Simulation results

Consider the simplified model of a vehicle in the form (5.3) given as (Guo et al., 2010; Yan et al., 2017)

$$\frac{\mathrm{d}}{\mathrm{d}t}\begin{bmatrix}\dot{\phi}\\ \dot{\xi}\\ \phi\\ \xi\end{bmatrix}=\begin{bmatrix}-2.11 & -6.61 & 9.48 & -357.05\\ 73.54 & -61.70 & 11.71 & -757.81\\ 1 & 0 & 0 & 0\\ 0 & 1 & 0 & 0\end{bmatrix}\begin{bmatrix}\dot{\phi}\\ \dot{\xi}\\ \phi\\ \xi\end{bmatrix}+\begin{bmatrix}1.2\\ 10\\ 0\\ 0\end{bmatrix}u+\begin{bmatrix}0\\ 8\\ 0\\ 0\end{bmatrix}d,$$

where ϕ and ξ denote the lean rotation and rotation of the front wheel with respect to the rear wheel, respectively. We will train up to level-5 intelligence. The system will be under attack by five agents with level 3, three agents with level 1, and two agents of level 5. Fig. 5.2 shows the convergence of the parameter λ after enough data have been gathered by interacting with the adversaries. After the parameter has converged, the defender's beliefs, described by (5.34), have taken the values shown in Fig. 5.3. In Fig. 5.4 the state exploration during the learning phase for the level-3 policy is shown, whereas the corresponding critic weights are shown in Fig. 5.5.

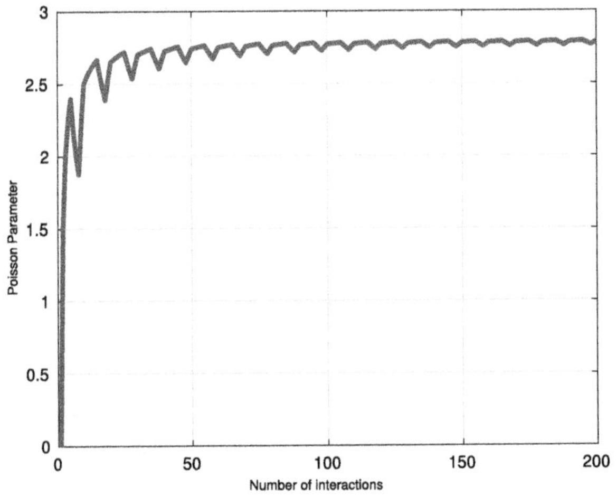

Figure 5.2 The Poisson parameter λ under the update rule (5.35).

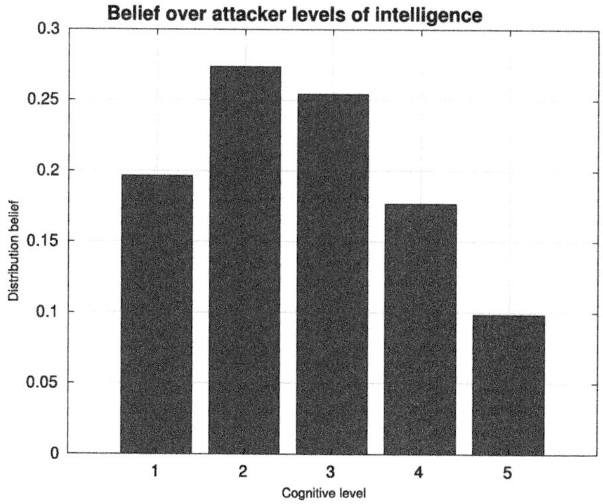

Figure 5.3 As long as the Poisson parameter has converged, the defender's belief over the levels of intelligence of the adversaries is biased toward the true levels of intelligence.

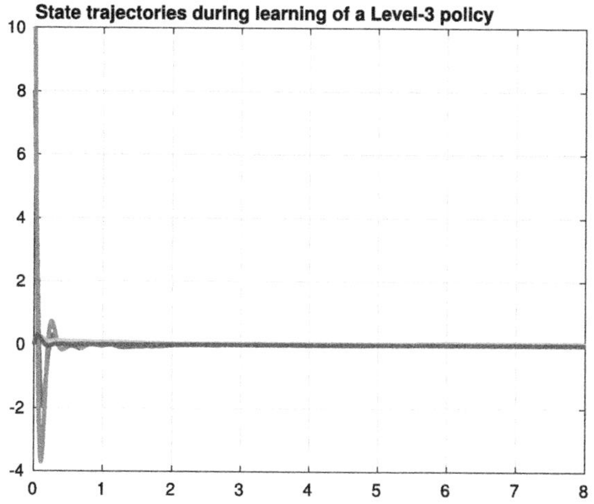

Figure 5.4 Evolution of the state trajectories during the learning process of the level-3 policy of the defender.

5.2. Non-equilibrium MDPs

We now formulate and solve non-equilibrium games in the context of stochastic games.

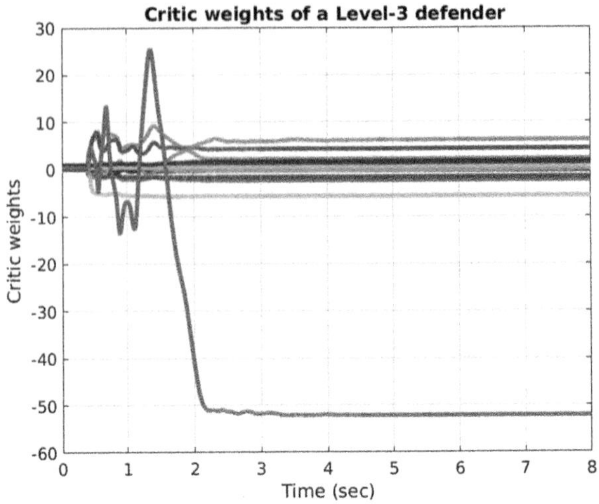

Figure 5.5 The evolution of the weights of the critic during the learning phase of a level-3 defense policy.

5.2.1 Problem formulation

Consider an N-player (agent) stochastic game defined by the tuple $(\mathcal{S},\ \mathcal{N},\ \mathcal{A},\ r,\ p,\ \gamma)$, where $\mathcal{S} = \{1, \ldots, |\mathcal{S}|\}$ is a finite state space, $\mathcal{N} = \{1, \ldots, N\}$ is a set of players, $\mathcal{A} = \mathcal{A}^1 \times \mathcal{A}^2 \times \cdots \times \mathcal{A}^N$ is a joint action space, with $\mathcal{A}^i$ being the individual finite action space of player $i \in \mathcal{N}$, $r = \{r^1, \ldots,\ r^N\}$ is a joint immediate reward function with $r^i : \mathcal{S} \times \mathcal{A}^i \times \mathcal{A}^{-i} \times \mathcal{S} \rightarrow \mathbb{R}^1$ being the individual immediate reward function of each player $i \in \mathcal{N}$, $p : \mathcal{S} \times \mathcal{A} \times \mathcal{S} \rightarrow [0,\ 1]$ is the conditional probability transition function, so that $p(s', a^1, \ldots, a^N, s)$ is the probability of transition from a state $s \in \mathcal{S}$ to a state $s' \in \mathcal{S}$ given actions $a^i \in \mathcal{A}^i$, $i \in \mathcal{N}$, and $\gamma \in (0, 1)$ is a discount factor.

For brevity, we also denote by $p^i(s', a^i, a^{-i}, s)$ the probability of transition from a state $s \in \mathcal{S}$ to a state $s' \in \mathcal{S}$ given actions $a^i \in \mathcal{A}^i$ and $a^{-i} \in \mathcal{A}^{-i}$, $i \in \mathcal{N}$. In addition, for $i \in \mathcal{N}$, we denote by $\mathcal{M}^i$ and $\mathcal{M}^{-i}$ the spaces of mappings $\mathcal{S} \rightarrow \mathcal{A}^i$ and $\mathcal{S} \rightarrow \mathcal{A}^{-i}$, respectively, and by $\mathcal{J}$ the space of mappings $\mathcal{S} \rightarrow \mathbb{R}$.

Given the stochastic game, each agent $i \in \mathcal{N}$ interacts over time with the environment and the other agents and takes an action $a_t^i \in \mathcal{A}^i$ at every time instant $t \in \mathbb{N}$. Owing to those actions, a sequence $\{s_t\}_{t \in \mathbb{N}}$ of states $s_t \in \mathcal{S}$ will be visited at each time instant $t \in \mathbb{N}$, depending on the conditional

[1] We denote $\mathcal{A}^{-i} = \times_{j \in \mathcal{N} \setminus \{i\}} \mathcal{A}^j$, $i \in \mathcal{N}$.

transition probabilities given by p and the initial state s_0. In this context, the goal of the player $i \in \mathcal{N}$ is to choose a policy $\pi^i \in \mathcal{M}^i$, i.e., a mapping describing which action $a^i \in \mathcal{A}^i$ is taken at any state $s \in \mathcal{S}$, so as to maximize their expected discounted cumulative reward, or value, given by

$$J^i_{\pi^i,\pi^{-i}}(s)=\mathbb{E}_p\left[\sum_{t=0}^{\infty}\gamma^t r^i(s_{t+1},\pi^i(s_t),\pi^{-i}(s_t),s_t)\;\middle|\; s_0=s\right], \tag{5.36}$$

where $\pi^{-i}=\{\pi^j\}_{j\in\mathcal{N}\setminus\{i\}}\in\mathcal{M}^{-i}$, and the expected value operator $\mathbb{E}_p$ is taken over the transition probabilities defined by p.

The optimal policy $\pi^{i\star}\in\mathcal{M}^i$ of player i, which maximizes (5.36), can be obtained through the following set of optimization problems:

$$\pi^{i\star}(s)\in\arg\max_{a^i\in\mathcal{A}^i}\sum_{s'\in\mathcal{S}}p^i\left(s',a^i,\pi^{-i}(s),s\right)\cdot\left(r^i\left(s,a^i,\pi^{-i}(s),s'\right)+\gamma J^i_{\pi^{i\star},\pi^{-i}}(s')\right),\quad s\in\mathcal{S},$$

where $J^i_{\pi^{i\star},\pi^{-i}}\in\mathcal{J}$ is the optimal value, which satisfies the Bellman equation

$$J^i_{\pi^{i\star},\pi^{-i}}(s)=\max_{a^i\in\mathcal{A}^i}\sum_{s'\in\mathcal{S}}p^i\left(s',a^i,\pi^{-i}(s),s\right)\cdot\left(r^i\left(s,a^i,\pi^{-i}(s),s'\right)+\gamma J^i_{\pi^{i\star},\pi^{-i}}(s')\right),\quad s\in\mathcal{S}. \tag{5.37}$$

Given (5.36)–(5.37), it is evident that the value of agent $i\in\mathcal{N}$ does not depend only on their policy, but also on the other agents' policies π^{-i}. These policies are generally unknown, and thus it is not straightforward for agent i to maximize (5.36); a model of the policies π^{-i} of the players in $\mathcal{N}\setminus\{i\}$ is needed for the maximization to be performed.

A common solution to this problem of lack of knowledge is for each agent $i\in\mathcal{N}$ to assume that all agents in $\mathcal{N}\setminus\{i\}$ will also optimize their own values and that this assumption is made by all agents. Hence, in this case, finding a policy that maximizes (5.36) is ultimately equivalent to computing a Nash equilibrium (Hespanha, 2017) according to the following definition.

Definition 5.2. The tuple $\{\mu^{i\star},\ \mu^{-i\star}\}$, $i\in\mathcal{N}$, with $\mu^{i\star}\in\mathcal{M}^i$ and $\mu^{-i\star}=\{\mu^{j\star}\}_{j\in\mathcal{N}\setminus\{i\}}\in\mathcal{M}^{-i}$, constitutes a Nash equilibrium if for all $\mu^i\in\mathcal{M}^i$ and $i\in\mathcal{N}$, we have

$$J^i_{\mu^{i\star},\mu^{-i\star}}(s)\geq J^i_{\mu^i,\mu^{-i\star}}(s),\quad s\in\mathcal{S}.\quad\blacksquare$$

Adopting the approach of the Nash equilibrium to model other agents' behaviors leads to two important issues. First, due to the finite nature of the action set $\mathcal{A}$, finding a Nash equilibrium requires a mostly intractable amount of computations even if all agents share the same reward functions r^i, $i \in \mathcal{N}$ (Bertsekas, 2020); second, in a realistic scenario, it is not necessary that all agents are perfectly rational,[2] and hence they may not operate on a Nash equilibrium (Camerer et al., 2004b), especially during initial plays of the learning mechanisms. Therefore, instead of relying on the concept of equilibrium, we will seek bounded rationality models to capture other agent behaviors. To this end, we use ideas from Camerer et al. (2004b); Strzalecki (2014) and using recursive reasoning, model different levels of rationality for each agent $i \in \mathcal{N}$ participating in the stochastic game.

5.2.2 Level-*k* thinking

Level-*k* thinking is a model of bounded rationality used to represent a player's strategy while also relaxing the assumption that every agent seeks a policy that is based solely on the notion of the Nash equilibrium. In particular, level-*k* thinking defines different levels of rationality, where at each level an agent assumes that the rest of the players follow a policy given by an immediately lower level. Then the agent proceeds to optimize their cumulative reward (5.36) given such an assumption. Taking the aforementioned into account, we can formulate a level-*k* thinking model for a stochastic game as follows.

Level-0: An agent $i \in \mathcal{N}$ with rationality of level-0, also defined as a level-0 agent, is a player that behaves naively; such an agent neither considers a model of the other agents' behavior nor tries to maximize their own cumulative reward (5.36). This is in line with Camerer (2003); Strzalecki (2014), which imposes the constraint that level-0 players are agnostic. Hence the policy $\pi_0^{i\star} \in \mathcal{M}^i$ of a level-0 agent $i \in \mathcal{N}$ can be chosen arbitrarily, so that

$$\pi_0^{i\star}(s) = a^i, \;\; a^i \in \mathcal{A}^i, \;\; s \in \mathcal{S}. \tag{5.38}$$

Note that there exist different ways to define a naive level-0 policy. Apart from choosing it as a constant one, the level-0 policy can also be chosen to be uniformly random (Strzalecki, 2014). In that case, $\pi_0^{i\star}(s) \sim \mathcal{U}\{\mathcal{A}^i\}$, $s \in \mathcal{S}$, with $\mathcal{U}\{\mathcal{A}^i\}$ being the uniform distribution over $\mathcal{A}^i$.

[2] In the sense that not every agent may be able to find the Nash equilibrium, that not every agent assumes that the rest of the agents will optimize their own values, or that not every agent actually seeks a Nash equilibrium.

Level-$k \in \mathbb{N}_+$: Unlike a level-0 agent, an agent $i \in \mathcal{N}$ with a rationality of level-k, $k \in \mathbb{N}_+$, reasons about the behavior of the other agents. In particular, for $k \in \mathbb{N}_+$, a level-k agent assumes that the level of rationality of the rest of the agents is $k-1$. Based on this assumption, a level-k agent acts strategically by trying to maximize their expected discounted cumulative reward and by choosing a level-k policy $\pi_k^{i\star} \in \mathcal{M}^i$ that satisfies, for all $k \in \mathbb{N}_+$,

$$\pi_k^{i\star} \in \arg\max_{\pi^i \in \mathcal{M}^i} J^i_{\pi^i,\ \pi_{k-1}^{-i\star}}(s), \quad s \in \mathcal{S}, \tag{5.39}$$

where $\pi_{k-1}^{-i\star} = \{\pi_{k-1}^{j\star}\}_{j \in \mathcal{N}\setminus\{i\}} \in \mathcal{M}^{-i}$. Since the action space $\mathcal{A}$ and the state space $\mathcal{S}$ are finite, there exists at least one policy satisfying (5.39). Hence, for any agent $i \in \mathcal{N}$, it is necessary and sufficient for a level-k policy $\pi_k^{i\star}$ to satisfy

$$J^i_{\pi_k^{i\star},\ \pi_{k-1}^{-i\star}}(s) \geq J^i_{\pi^i,\ \pi_{k-1}^{-i\star}}(s), \quad s \in \mathcal{S}, \ \pi^i \in \mathcal{M}^i.$$

5.2.3 Level-recursive computation of level-k policies

We proceed to find the level-k policies described by (5.39) for all $k \in \mathbb{N}_+$. To this end, we define the Bellman operator $T^i_{\mu^{-i}}$ and the Q-factor operators $Q^i_{\mu^{-i}}$, $i \in \mathcal{N}$, that map functions of the form $J \in \mathcal{J}$ to functions of the form

$$\left(T^i_{\mu^{-i}}J\right)(s) := \max_{a^i \in \mathcal{A}^i} \sum_{s' \in \mathcal{S}} p^i\left(s', a^i, \mu^{-i}(s), s\right) \cdot \left(r^i\left(s, a^i, \mu^{-i}(s), s'\right) + \gamma J(s')\right), \quad s \in \mathcal{S}, \tag{5.40}$$

and

$$\left(Q^i_{\mu^{-i}}J\right)(s, a^i) := \sum_{s' \in \mathcal{S}} p^i\left(s', a^i, \mu^{-i}(s), s\right) \cdot \left(r^i\left(s, a^i, \mu^{-i}(s), s'\right) + \gamma J(s')\right), \quad s \in \mathcal{S}, \ a^i \in \mathcal{A}^i, \tag{5.41}$$

for any policies $\mu^{-i} \in \mathcal{M}^{-i}$.

The aforementioned operators have the following properties.

Lemma 5.2. *Let $\pi_k^{i\star} \in \mathcal{M}^i$ be a level-k policy defined in* (5.38)–(5.39) *with $k \in \mathbb{N}$ and $i \in \mathcal{N}$. Then, for $k \in \mathbb{N}_+$, the equation*

$$T^i_{\pi_{k-1}^{-i\star}} J(s) = J(s), \quad s \in \mathcal{S}, \tag{5.42}$$

has a unique fixed point with respect to $J \in \mathcal{J}$ given by the value $J^i_{\pi_k^{i\star},\ \pi_{k-1}^{-i\star}}$.

Proof. The proof follows from the fact that (5.42) is a Bellman equation (Bertsekas and Tsitsiklis, 1996) with a fixed point given by $J^i_{\pi^{i\star}_k,\ \pi^{-i\star}_{k-1}}$ since $\pi^{i\star}_k \in \arg\max_{\pi^i \in \mathcal{M}^i} J^i_{\pi^i,\ \pi^{-i\star}_{k-1}}(s)$, $s \in \mathcal{S}$, $k \in \mathbb{N}_+$. □

Lemma 5.3. *Let $\pi^{i\star}_k \in \mathcal{M}^i$ be a level-k policy defined as in (5.38)–(5.39) for all $k \in \mathbb{N}$ and $i \in \mathcal{N}$. Then the Bellman operator $T^i_{\pi^{-i\star}_{k-1}}$, where $k \in \mathbb{N}_+$ and $i \in \mathcal{N}$, is a contraction mapping in the sense that*

$$\left\| T^i_{\pi^{-i\star}_{k-1}} J - T^i_{\pi^{-i\star}_{k-1}} J' \right\|_\infty \leq \gamma \left\| J - J' \right\|_\infty$$

for all $J,\ J' \in \mathcal{J}$.

Proof. The proof follows from Bertsekas and Tsitsiklis (1996), owing to the fact that $\gamma \in (0, 1)$ and due to the finiteness of the state space $\mathcal{S}$ and the action space $\mathcal{A}$. □

Using Lemmas 5.2–5.3, the following theorem defines a level-recursive procedure that provably computes a set of level-k policies.

Theorem 5.3. *Let $\hat{\pi}^i_k \in \mathcal{M}^i$ be a policy estimate of a level-k policy $\pi^{i\star}_k$ described by (5.38)–(5.39) for all $i \in \mathcal{N}$ and $k \in \mathbb{N}$. For $k = 0$, assume that the level-0 policy estimate is given by*

$$\hat{\pi}^i_0 := \pi^{i\star}_0, \quad i \in \mathcal{N}. \tag{5.43}$$

In addition, for all $i \in \mathcal{N}$, define the following recursive procedure on $k \in \{1, \ldots, k^\star\}$, where $k^\star \in \mathbb{N}_+$:

$$\begin{aligned} J^i_{k,0}(s) &\text{ randomized}, \quad s \in \mathcal{S}, \\ J^i_{k,n+1}(s) &= \left(T^i_{\hat{\pi}^{-i}_{k-1}} J^i_{k,n} \right)(s), \quad s \in \mathcal{S}, \quad n \in \mathbb{N}, \\ \hat{\pi}^i_k(s) &\in \arg\max_{a^i \in \mathcal{A}^i} \left(Q^i_{\hat{\pi}^{-i}_{k-1}} J^i_{k,n^\star} \right)(s, a^i), \quad s \in \mathcal{S}, \end{aligned} \tag{5.44}$$

where $\hat{\pi}^{-i}_{k-1} = \{\hat{\pi}^j_{k-1}\}_{j \in \mathcal{N}\setminus\{i\}}$, $\{J^i_{k,n}\}_{n\in\mathbb{N}}$ is a sequence of functions $J^i_{k,n} \in \mathcal{J}$, and $n^\star > 0$ is a sufficiently large integer. Then $\hat{\pi}^i_k \equiv \pi^{i\star}_k$, $i \in \mathcal{N}$, $k \in \{0, \ldots, k^\star\}$.

Proof. The proof follows a "step-based" procedure, where at each step $k = \{0, \ldots, k^\star\}$, it is proved that $\hat{\pi}^i_k(s) = \pi^{i\star}_k(s)$, $s \in \mathcal{S}$, $i \in \mathcal{N}$, for some level-k policy $\pi^{i\star}_k$ satisfying (5.38)–(5.39).

Step 0: The relation $\hat{\pi}^i_0 \equiv \pi^{i\star}_0$ holds due to (5.43), since $\pi^{i\star}_0$ is known a priori from (5.38).

Step 1: Since $\hat{\pi}_0^i \equiv \pi_0^{i\star}$, $i \in \mathcal{N}$, for $k = 1$, iteration (5.44) reduces to

$$J_{1,n+1}^i(s) = \left(T_{\pi_0^{-i\star}}^i J_{1,n}^i\right)(s), \quad s \in \mathcal{S}, \quad n \in \mathbb{N}. \tag{5.45}$$

Due to (5.45), we use the results of Lemmas 5.2 and 5.3 to conclude that

$$\lim_{n\to\infty} J_{1,n}^i(s) = J_{\pi_1^{i\star},\ \pi_0^{-i\star}}^i(s), \quad s \in \mathcal{S}.$$

Hence, for all $\epsilon > 0$, there exists a positive integer n_1 such that $\left\| J_{1,n}^i - J_{\pi_1^{i\star},\ \pi_0^{-i\star}}^i \right\|_\infty < \epsilon$ for all $n \geq n_1$. In that respect, owing to the finite action space $\mathcal{A}$ and state space $\mathcal{S}$, there also exists a positive integer n_2 such that $\arg\max_{a^i \in \mathcal{A}^i}(Q_{\pi_0^{-i\star}}^i J_{1,n}^i)(s, a^i) \subseteq \arg\max_{a^i \in \mathcal{A}^i}(Q_{\pi_0^{-i\star}}^i J_{\pi_1^{i\star},\ \pi_0^{-i\star}}^i)(s, a^i)$ for all $n \geq n_2$. Hence, if $n^\star \geq n_2$, then by (5.44)

$$\hat{\pi}_1^i(s) \in \arg\max_{a^i \in \mathcal{A}^i}(Q_{\pi_0^{-i\star}}^i J_{\pi_1^{i\star},\ \pi_0^{-i\star}}^i)(s, a^i), \quad s \in \mathcal{S}. \tag{5.46}$$

However, due to Lemma 5.2, $\max_{a^i \in \mathcal{A}^i}(Q_{\pi_0^{-i\star}}^i J_{\pi_1^{i\star}, \pi_0^{-i\star}}^i)(s, a^i) = T_{\pi_0^{-i\star}}^i J_{\pi_1^{i\star}, \pi_0^{-i\star}}^i(s) = J_{\pi_1^{i\star},\ \pi_0^{-i\star}}^i(s)$, $s \in \mathcal{S}$, and hence from (5.39) and (5.46) it follows that $\hat{\pi}_1^i \equiv \pi_1^{i\star}$.

Step $k \in \{2, \ldots, k^\star\}$**:** The results of Step 1 can be generalized inductively to conclude $\hat{\pi}_k^i \equiv \pi_k^{i\star}$, $i \in \mathcal{N}$. □

A procedure summarizing the practical implementation of Theorem 5.3 is provided in Algorithm 5.2.

5.2.4 Level-paralleled computation of level-*k* policies

The recursion presented in Algorithm 5.2, which can be used to solve (5.39), entails a computational hurdle; in order for the execution of step $k \in \mathbb{N}_+$ of the recursion (5.44) to begin, the previous step $k - 1$ needs to have terminated. This might be inefficient for learning mechanisms because any information generated during the execution of step $k - 1$ cannot be immediately used by step k. Hence we may desire to implement a version of Algorithm 5.2 that updates the value estimates J_k^i and the policy estimates $\hat{\pi}_k^i$ simultaneously over all $k \in \mathbb{N}_+$. Such a procedure is described in Algorithm 5.3.

Although Algorithm 5.3 allows for the parallel estimation of the level-k policies over all the levels $k \in \mathbb{N}$, its convergence relies on the following uniqueness assumption.

Algorithm 5.2 Level-Recursive Computation of the Level-k Policies.

Input: Sufficiently small constant $\epsilon > 0$, maximum level $k^\star \in \mathbb{N}_+$.
Output: Level-k policies $\hat{\pi}_k^i = \pi_k^{i\star}$, $\forall i \in \mathcal{N}$, $\forall k \in \{0, \ldots, k^\star\}$.

procedure
  for $i = 1, \ldots, N$ **do** ▷ Initialization
    $\hat{\pi}_0^i(s) \leftarrow \pi_0^{i\star}(s), \ \forall s \in \mathcal{S}$.
    for $k = 1, \ldots, k^\star$ **do**
      Initialize $J_k^i(s)$ randomly, $\forall s \in \mathcal{S}$.
    end for
  end for
  for $k = 1, \ldots, k^\star$ **do** ▷ Level-k policy estimation
    for $i = 1, \ldots, N$ **do**
      repeat
        $v(s) \leftarrow J_k^i(s), \ \forall s \in \mathcal{S}$.
        $J_k^i(s) \leftarrow T^i_{\hat{\pi}_{k-1}^{-i}} v(s), \ \forall s \in \mathcal{S}$.
      until $\left\| J_k^i - v \right\|_\infty < \epsilon$.
      $\hat{\pi}_k^i(s) \leftarrow \text{argmax}_{a^i \in \mathcal{A}^i} (Q^i_{\hat{\pi}_{k-1}^{-i}} J_k^i)(s, a^i), \ \forall s \in \mathcal{S}$.
    end for
  end for
end procedure

Assumption 5.4. The cost functions of any two distinct policies are distinct, i.e., for any two policies $\mu^i, \ \mu'^i \in \mathcal{M}^i$ and a joint policy $\mu^{-i} \in \mathcal{M}^{-i}$, $i \in \mathcal{N}$, we have

$$\mu^i \neq \mu'^i \Longrightarrow J^i_{\mu^i, \ \mu^{-i}} \neq J^i_{\mu'^i, \ \mu^{-i}}. \quad \blacksquare$$

Remark 5.7. Although Assumption 5.4 restricts the use of Algorithm 5.3 to a particular class of stochastic games, it is commonly imposed in the literature when value iteration is used in a multi-agent framework (Bertsekas, 2020). ■

The following lemmas are critical in establishing the convergence properties of Algorithm 5.3. In particular, Lemma 5.4 shows that Algorithm 5.3 will terminate only with the level-k policies as its output, whereas Lemma 5.5 proves that Algorithm 5.3 will indeed terminate. Theorem 5.4 combines these two results to conclude the convergence of Algorithm 5.3.

Algorithm 5.3 Level-Paralleled Computation of the Level-k Policies.

Input: Sufficiently small constant $\epsilon > 0$, maximum level $k^\star \in \mathbb{N}_+$.
Output: Level-k policies $\hat{\pi}^i_k = \pi^{i\star}_k$, $\forall i \in \mathcal{N}$, $\forall k \in \{0, \ldots, k^\star\}$.

1: **procedure**
2:     **for** $i = 1, \ldots, N$ **do** ▷ Initialization
3:         $\hat{\pi}^i_0(s) \leftarrow \pi^{i\star}_0(s), \ \forall s \in \mathcal{S}$.
4:         **for** $k = 1, \ldots, k^\star$ **do**
5:             Initialize $J^i_k(s)$ randomly, $\forall s \in \mathcal{S}$.
6:         **end for**
7:     **end for**
8:     **repeat** ▷ Level-k policy estimation
9:         $\Delta \leftarrow 0$.
10:         **for** $k = 1, \ldots, k^\star$ **do**
11:             **for** $i = 1, \ldots, N$ **do**
12:                 $v(s) \leftarrow J^i_k(s), \ \forall s \in \mathcal{S}$.
13:                 $J^i_k(s) \leftarrow T^i_{\hat{\pi}^{-i}_{k-1}} v(s), \ \forall s \in \mathcal{S}$.
14:                 $\hat{\pi}^i_k(s) \leftarrow \arg\max_{a^i \in \mathcal{A}^i} (Q^i_{\hat{\pi}^{-i}_{k-1}} J^i_k)(s, a^i), \ \forall s \in \mathcal{S}$.
15:                 $\Delta \leftarrow \max\left\{\Delta, \ \left\|J^i_k - v\right\|_\infty\right\}$.
16:             **end for**
17:         **end for**
18:     **until** $\Delta < \epsilon$.
19: **end procedure**

Lemma 5.4. *Let Assumption* 5.4 *hold. Let* $\hat{\pi}^i_k \in \mathcal{M}^i$ *for all* $i \in \mathcal{N}$ *and* $k = \{0, \ldots, k^\star\}$ *be the output of Algorithm* 5.3, $k^\star \in \mathbb{N}_+$ *and* $\epsilon > 0$ *given as inputs. Given that* $\epsilon > 0$ *is sufficiently small, Algorithm* 5.3 *will terminate only if* $\hat{\pi}^i_k = \pi^{i\star}_k$ *for all* $i \in \mathcal{N}$ *and* $k = \{0, \ldots, k^\star\}$.

Proof. Let us assume that there exists $\bar{k} \in \{1, \ldots, k^\star\}$ such that $\hat{\pi}^i_{\bar{k}-1} = \pi^{i\star}_{\bar{k}-1}$ for all $i \in \mathcal{N}$ at an iteration of the loop in lines 8–18 of Algorithm 5.3. Then Algorithm 5.3 will terminate at that iteration only if

$$\left\| J^i_{\bar{k}} - T^i_{\hat{\pi}^{-i}_{\bar{k}-1}} J^i_{\bar{k}} \right\|_\infty < \epsilon \Rightarrow \left\| J^i_{\bar{k}} - T^i_{\pi^{-i\star}_{\bar{k}-1}} J^i_{\bar{k}} \right\|_\infty < \epsilon \tag{5.47}$$

for all $i \in \mathcal{N}$. However, according to Lemma 5.2, $J^i_{\pi^{i\star}_{\bar{k}}, \pi^{-i\star}_{\bar{k}-1}}$ is the unique fixed point of the operator $T^i_{\pi^{-i\star}_{\bar{k}-1}}$, and hence the necessary condition for termination (5.47) is equivalent to

$$\left\| J^i_{\bar{k}} - J^i_{\pi^{i\star}_{\bar{k}},\, \pi^{-i\star}_{\bar{k}-1}} \right\|_\infty < \phi(\epsilon), \ i \in \mathcal{N}, \tag{5.48}$$

where $\phi(\epsilon)>0$ and $\phi(\epsilon)\to 0$ as $\epsilon\to 0$. Hence, given a sufficiently small $\epsilon>0$, due to Assumption 5.4 and (5.48), Algorithm 5.3 will terminate at that iteration only if $\arg\max_{a^i\in\mathcal{A}^i}(Q^i_{\pi^{-i\star}_{\bar{k}-1}} J^i_{\bar{k}})(s, a^i) = \arg\max_{a^i\in\mathcal{A}^i}(Q^i_{\pi^{-i\star}_{\bar{k}-1}} J^i_{\pi^{i\star}_{\bar{k}},\, \pi^{-i\star}_{\bar{k}-1}})(s, a^i)$, $s \in \mathcal{S}$, $i \in \mathcal{N}$. Since, for all $s \in \mathcal{S}$ and $i \in \mathcal{N}$, $\max_{a^i\in\mathcal{A}^i}(Q^i_{\pi^{-i\star}_{\bar{k}-1}} J^i_{\pi^{i\star}_{\bar{k}},\, \pi^{-i\star}_{\bar{k}-1}})(s, a^i)$ $= T^i_{\pi^{-i\star}_{\bar{k}-1}} J^i_{\pi^{i\star}_{\bar{k}},\, \pi^{-i\star}_{\bar{k}-1}}(s) = J^i_{\pi^{i\star}_{\bar{k}},\, \pi^{-i\star}_{\bar{k}-1}}(s)$, Algorithm 5.3 will terminate at that iteration only if $\hat{\pi}^i_{\bar{k}} = \pi^{i\star}_{\bar{k}}$ for all $i \in \mathcal{N}$, owing to (5.39).

We have shown that if there exists $\bar{k} \in \{1, \dots, k^\star\}$ such that $\hat{\pi}^i_{\bar{k}-1} = \pi^{i\star}_{\bar{k}-1}$ is true for all $i \in \mathcal{N}$ during an iteration of the loop in lines 8–18 of Algorithm 5.3, then Algorithm 5.3 will terminate only if $\hat{\pi}^i_{\bar{k}} = \pi^{i\star}_{\bar{k}}$ at that iteration for all $i \in \mathcal{N}$. To exploit this property, note that $\hat{\pi}^i_0 = \pi^{i\star}_0$ is always true for all $i \in \mathcal{N}$. Hence Algorithm 5.3 will not terminate if there exists $i \in \mathcal{N}$ such that $\hat{\pi}^i_1 \neq \pi^{i\star}_1$. In addition, if $\hat{\pi}^i_1 = \pi^{i\star}_1$ for all $i \in \mathcal{N}$, then Algorithm 5.3 will not terminate if there exists $i \in \mathcal{N}$ such that $\hat{\pi}^i_2 \neq \pi^{i\star}_2$. Generalizing inductively, we conclude that Algorithm 5.3 will terminate only if $\hat{\pi}^i_k = \pi^{i\star}_k$ for all $i \in \mathcal{N}$ and $k \in \{0, \dots, k^\star\}$. □

The following lemma now provides sufficient conditions for termination, which complement the foregoing necessary conditions.

Lemma 5.5. *Let Assumption* 2.1 *hold, and assume that* $\epsilon > 0$ *and* $k^\star \in \mathbb{N}_+$ *are the inputs of Algorithm* 5.3. *Then Algorithm* 5.3 *will terminate.*

Proof. Let us assume, ad absurdum, that Algorithm 5.3 will not terminate. Then we can denote as $\{J^i_{k,n}\}$ and $\{\hat{\pi}^i_{k,n}\}$ the infinite sequences of J^i_k and $\hat{\pi}^i_k$ that will be generated at each iteration $n \in \mathbb{N}$ of the loop in lines 8–18 of Algorithm 5.3, $i \in \mathcal{N}$, $k \in \{0, \dots, k^\star\}$. We now follow a $(k^\star + 1)$-step reasoning procedure to arrive at a contradiction.

Step 0**:** The relation $\hat{\pi}^i_0 = \pi^{i\star}_0$ holds trivially due to line 3 of Algorithm 5.3 and since $\pi^{i\star}_0$ is known a priori. Hence $\hat{\pi}^i_{0,n} = \pi^{i\star}_0$ for all $n \geq 0$ and $i \in \mathcal{N}$.

Step $k \in \{1, \dots, k^\star\}$: Assume that there exists a non-negative integer $N^{i\star}_{k-1}$ such that $\hat{\pi}^i_{k-1,n} = \pi^{i\star}_{k-1}$, $n \geq N^{i\star}_{k-1}$, $i \in \mathcal{N}$. We will now prove by induction that there also exists a non-negative integer $N^{i\star}_k$ such that $\hat{\pi}^i_{k,n} = \pi^{i\star}_k$, $n \geq N^{i\star}_k$, $i \in \mathcal{N}$.

Algorithm 5.3 executes, for all $n \geq \max_{i\in\mathcal{N}} N_{k-1}^{i\star} =: \bar{N}_{k-1}^{\star}$ and $i \in \mathcal{N}$, the iteration

$$J_{k,n+1}^{i}(s) = T_{\pi_{k-1}^{-i\star}} J_{k,n}^{i}(s), \quad s \in \mathcal{S}.$$

Hence, after using the results of Lemmas 5.2 and 5.3, we conclude that for $n \geq \bar{N}_{k-1}^{\star}$ and $i \in \mathcal{N}$,

$$\left\| J_{k,n}^{i} - J_{\pi_k^{i\star},\, \pi_{k-1}^{-i\star}}^{i} \right\|_{\infty} \leq \gamma^{n-\bar{N}_{k-1}^{\star}} \left\| J_{k,N_{k-1}^{\star}}^{i} - J_{\pi_k^{i\star},\, \pi_{k-1}^{-i\star}}^{i} \right\|_{\infty}. \tag{5.49}$$

Due to (5.49), for all $i \in \mathcal{N}$ and $\delta_k^i > 0$, there exists a bounded constant $n_k^{i\star}(\delta_k^i) > 0$ such that

$$\left\| J_{k,n}^{i} - J_{\pi_k^{i\star},\, \pi_{k-1}^{-i\star}}^{i} \right\|_{\infty} < \delta_k^i, \ \forall n \geq n_k^{i\star}(\delta_k^i).$$

Hence, for all $i \in \mathcal{N}$, after taking into account Assumption 1, the finite action space $\mathcal{A}$, and the finite state space $\mathcal{S}$, there exists a non-negative integer $N_k^{i\star} \geq \bar{N}_{k-1}^{\star}$ such that $\arg\max_{a^i\in\mathcal{A}^i}(Q_{\pi_{k-1}^{-i\star}}^{i} J_{k,n}^{i})(s, a^i) = \arg\max_{a^i\in\mathcal{A}^i}(Q_{\pi_{k-1}^{-i\star}}^{i} J_{\pi_k^{i\star},\, \pi_{k-1}^{-i\star}}^{i})(s, a^i)$ for all $n \geq N_k^{i\star}$. However, given Lemma 5.2, $\max_{a^i\in\mathcal{A}^i}(Q_{\pi_{k-1}^{-i\star}}^{i} J_{\pi_k^{i\star},\, \pi_{k-1}^{-i\star}}^{i})(s, a^i) = T_{\pi_{k-1}^{-i\star}}^{i} J_{\pi_k^{i\star},\, \pi_{k-1}^{-i\star}}^{i}(s) = J_{\pi_k^{i\star},\, \pi_{k-1}^{-i\star}}^{i}(s)$. Therefore, due to line 14 of Algorithm 5.3 and (5.39),

$$\hat{\pi}_{k,n}^{i} = \pi_k^{i\star}, \ n \geq N_k^{i\star}, \ i \in \mathcal{N}. \tag{5.50}$$

Hence the induction is concluded.

Finally, from (5.50), note that for $n \geq \max_{i\in\mathcal{N}} N_{k^\star}^{i\star}$, Algorithm 5.3 executes, for all $k \in \{1, \ldots, k^\star\}$, the iteration $J_{k,n+1}^{i}(s) = T_{\pi_{k-1}^{-i\star}}^{i} J_{k,n}^{i}(s)$, $s \in \mathcal{S}$, $i \in \mathcal{N}$. Since $T_{\pi_{k-1}^{-i\star}}^{i}$ is a contraction mapping, $\left\| J_{k,n+1}^{i} - J_{k,n}^{i} \right\|_{\infty} < \epsilon$ will hold eventually, $i \in \mathcal{N}$, $k \in \{1, \ldots, k^\star\}$. Hence Algorithm 5.3 will terminate, which contradicts the assumption we initially made. □

Theorem 5.4. *Suppose that Assumption* 2.1 *holds. Assume that* $\epsilon > 0$ *and* $k^\star \in \mathbb{N}_+$ *are the inputs of Algorithm* 5.3. *Then, given a sufficiently small* ϵ, *the output of Algorithm* 5.3 *is* $\hat{\pi}_k^i = \pi_k^{i\star}$, $k \in \{0, \ldots, k^\star\}$, $i \in \mathcal{N}$, *where* $\pi_k^{i\star}$ *is the level-k policy as defined in* (5.38)–(5.39).

Proof. Owing to Lemma 5.5, Algorithm 5.3 terminates. However, due to Lemma 5.4, Algorithm 5.3 terminates only if $\hat{\pi}_k^i = \pi_k^{i\star}$, $k \in \{0,\ldots,k^\star\}$, $i \in \mathcal{N}$. Hence the result follows. □

Remark 5.8. The "for-loop" in lines 10–17 of Algorithm 5.3 can be executed in a distributed/asynchronous manner (as in Bertsekas and Tsitsiklis, 2015) without having any effect on the convergence of the algorithm. In fact, several iterations of this loop can be omitted at multiple instances of the wider "repeat-while" loop in lines 8–18. Hence, given that the state space S is large, a significant speed-up can be attained, which is not possible with the level-recursive Algorithm 5.2. ■

Remark 5.9. In the level-paralleled Algorithm 5.3, we can assess whether each level k has *individually* converged with a margin of error ϵ. If such a convergence has been achieved, then the level can be removed from the for-loop as long as all lower levels have been removed from the for-loop in a previous round, after also having converged with a margin of error ϵ. In that case the uniqueness Assumption 5.4 will not be needed. ■

5.2.5 Cognitive hierarchy

According to the level-k thinking model presented previously, a level-k agent assumes that the rest of the agents are level-$(k-1)$ for all $k \in \mathbb{N}_+$. However, such an assumption can be restrictive; if the rest of the agents are at a lower level of rationality, but not exactly at $k-1$, the all optimality guarantees are dropped. Therefore it is of interest to construct a more generalized model of bounded rationality that allows for the other agents' levels to vary and not be deterministically equal to $k-1$.

To this end, to generalize level-k thinking, we construct a bounded rationality model based on *cognitive hierarchy* (Camerer et al., 2004b). According to this model, a level-k agent does not necessarily assume that the rest of the agents are level-$(k-1)$, but that their cognitive level follows a distribution over $\{0, 1, \ldots, k-1\}$. If g is a probability mass function over $\mathbb{N}$, then such a distribution $\mathcal{P}_k$ over $\kappa \in \{0, 1, \ldots, k-1\}$, $k \in \mathbb{N}_+$, can be defined by the probability mass function

$$P_k(\kappa) = \frac{g(\kappa)}{\sum_{i=0}^{k-1} g(i)}, \quad \kappa \in \{0, 1, \ldots, k-1\}. \tag{5.51}$$

It is common to select g to represent a Poisson distribution, since experiments have shown that the proportion of players with a cognitive level of $k-1$ usually decreases as k increases (Camerer et al., 2004a). By adopting the Poisson model we have

$$g(\kappa) = \frac{\lambda^\kappa e^{-\lambda}}{\kappa!}, \tag{5.52}$$

where $\lambda > 0$ is the mean and variance of the Poisson model.

Given (5.51)–(5.52), the cognitive hierarchy model derives the following policies $\mu_k^{i\star} \in \mathcal{M}^i$ at each level $k \in \mathbb{N}$ for all $i \in \mathcal{N}$.

Level 0: The level-0 policy in cognitive hierarchy is defined exactly as in the case of level-k thinking, that is,

$$\mu_0^{i\star}(s) = a^i, \ a^i \in \mathcal{A}^i, \ s \in \mathcal{S}.$$

Level $k \in \mathbb{N}_+$: According to the bounded rationality model of cognitive hierarchy, an agent $i \in \mathcal{N}$ of level-k, $k \in \mathbb{N}_+$, assumes that each of the other agents has a level of intelligence κ, given by distribution (5.51)–(5.52), that is, $\kappa \sim \mathcal{P}_k$. Since κ is a random variable, it is in the interest of agent $i \in \mathcal{N}$ to maximize the expectation of their value over $\kappa \sim \mathcal{P}_k$ and pick their policy according to

$$\mu_k^{i\star} \in \arg\max_{\mu^i \in \mathcal{M}^i} \mathbb{E}\left[J^i_{\mu^i,\ \mu_\kappa^{-i\star}}(s) \mid \kappa \sim \mathcal{P}_k\right], \ s \in \mathcal{S}, \tag{5.53}$$

where $\mu_\kappa^{-i\star} = \{\mu_\kappa^{j\star}\}_{j \in \mathcal{N}\setminus\{i\}}$. By slightly modifying operators (5.40)–(5.41) so that they are taken with respect to the expected value of the now random policies of the agents in $\mathcal{N}\setminus\{i\}$, it is straightforward to extend Algorithms 5.2 and 5.3 to solve (5.53). We omit the corresponding analysis due to space limitations.

Remark 5.10. The Poisson distribution defined by (5.52), which we used in the cognitive hierarchy model, follows from the experimental analysis and data (Camerer, 2003, 2016; Camerer et al., 2004b). In particular, the aforementioned studies have shown that cognitive hierarchy models, which are based on a Poisson distribution, can accurately predict the behavior of boundedly rational agents. Nevertheless, should one desire to define (5.52) in a different manner, and with a different domain, the convergence of Algorithms 5.2–5.3 would not be affected. ■

Cognitive hierarchy enjoys some convergence properties that are not inherent to level-k thinking. In particular, as the level of intelligence increases, the cost function and the corresponding policy will converge to some limit functions. This is illustrated in the following theorem.

Theorem 5.5. *Consider the cognitive hierarchy model described by* (5.51)–(5.53). *Then the sequence of the cognitive hierarchy value functions* $\{\tilde{J}_k^i(s)\}_{k \in \mathbb{N}}$, $s \in \mathcal{S}$, *where* $\tilde{J}_k^i \in \mathcal{J}$ *is defined as*

$$\tilde{J}_k^i(s) := \mathbb{E}\left[J^i_{\mu_k^{i\star},\ \mu_\kappa^{-i\star}}(s) \mid \kappa \sim \mathcal{P}_k\right],\ s \in \mathcal{S},\ k \in \mathbb{N}_+,$$

is convergent as $k \to \infty$.

Proof. Notice that for $\kappa \in \mathbb{N}$ and $s \in \mathcal{S}$, we have

$$\begin{aligned}\left|g(\kappa) J^i_{\mu_k^{i\star},\ \mu_\kappa^{-i\star}}(s)\right| &\leq \left|g(\kappa)\right| \left(\left|J^i_{\mu_k^{i\star},\ \mu_\kappa^{-i\star}}(s)\right| + 1\right) \\ &\leq \left|g(\kappa)\right| (\bar{J}^i + 1),\end{aligned} \tag{5.54}$$

where $\bar{J}^i \geq 0$ denotes the greatest possible upper bound of the absolute value of (5.36) for player $i \in \mathcal{N}$ and is finite. In particular, this bound is defined as

$$\bar{J}^i = \max_{s \in \mathcal{S},\ \pi^i \in \mathcal{M}^i,\ \pi^{-i} \in \mathcal{M}^{-i}} |J^i_{\pi^i, \pi^{-i}}(s)|.$$

Define now $c_\kappa := \left|g(\kappa)\right| (\bar{J}^i + 1)$, $\kappa \in \mathbb{N}$. Then it follows that

$$\lim_{\kappa \to \infty} \frac{|c_{\kappa+1}|}{|c_\kappa|} = \lim_{\kappa \to \infty} \frac{\left|g(\kappa+1)\right| (\bar{J}^i + 1)}{\left|g(\kappa)\right| (\bar{J}^i + 1)} = \lim_{\kappa \to \infty} \frac{\lambda}{\kappa + 1} = 0.$$

Therefore the series $\sum_{\kappa=0}^{\infty} c_\kappa$ is absolutely convergent (Apostol, 1991). In addition, if we let $b_\kappa = g(\kappa) J^i_{\mu_k^{i\star},\ \mu_\kappa^{-i\star}}(s)$, then the series $\sum_{\kappa=0}^{\infty} b_\kappa$ is also absolutely convergent since $|b_\kappa| \leq |c_\kappa|$ due to (5.54) and because $\sum_{\kappa=0}^{\infty} c_k$ has been proved to absolutely converge. Finally, since

$$\begin{aligned}\tilde{J}_k^i(s) &= \mathbb{E}\left[J^i_{\mu_k^{i\star},\ \mu_\kappa^{-i\star}}(s) \mid \kappa \sim \mathcal{P}_k\right] \\ &= \sum_{\kappa=0}^{k-1} P_k(\kappa) J^i_{\mu_k^{i\star},\ \mu_\kappa^{-i\star}}(s) = \frac{1}{\sum_{i=0}^{k-1} g(i)} \sum_{\kappa=0}^{k-1} b_\kappa,\end{aligned}$$

where

$$\lim_{k \to \infty} \sum_{i=0}^{k-1} g(i) = 1,$$

and

$$\lim_{k \to \infty} \sum_{\kappa=0}^{k-1} b_\kappa \text{ is finite,}$$

we conclude that $\{\tilde{J}_k^i(s)\}_{k \in \mathbb{N}}$ converges to a finite limit as $k \to \infty$. □

The cost function has been proved to converge as $k \to \infty$. The same will also hold for the corresponding policies, given a uniqueness condition in the same fashion as in Assumption 2.3.

In addition, note that Theorem 5.5 does not prove the convergence to a pure Nash equilibrium as $k \to \infty$, as such an equilibrium may not even exist. Nevertheless, this should not be considered a weakness of cognitive hierarchy, because the purpose of boundedly rational models is not to approximate the Nash equilibrium, but rather to complement it in scenarios where it is either inapplicable, hard to compute, or perhaps even inappropriate.

Still, in the literature of game theory (Camerer et al., 2004b), cognitive hierarchy does exhibit an interesting property with regards to Nash equilibria. Namely, given that the agents' policies form a Nash equilibrium at some level, then they will also form an equilibrium at all higher-order levels. In what follows, we will prove that this statement also holds for the cognitive hierarchy model of this section.

Theorem 5.6. *Consider the cognitive hierarchy model described by* (5.51)–(5.53). *Assume that, at some level $\bar{k} \in \mathbb{N}$, the corresponding level-$\bar{k}$ policy tuple $\{\mu_{\bar{k}}^{i\star},\ \mu_{\bar{k}}^{-i\star}\} \in \mathcal{M}^i \times \mathcal{M}^{-i}$ forms a Nash equilibrium for all $i \in \mathcal{N}$. Then the higher-level policy tuples $\{\mu_k^{i\star},\ \mu_k^{-i\star}\} \in \mathcal{M}^i \times \mathcal{M}^{-i}$, $k \in \mathbb{N}_{\geq \bar{k}}$, will also form Nash equilibria.*

Proof. Let the policy tuple $\{\mu_{\bar{k}}^{i\star},\ \mu_{\bar{k}}^{-i\star}\}$, at some level $\bar{k} \in \mathbb{N}$ of the cognitive hierarchy model and for all $i \in \mathcal{N}$, be a Nash equilibrium. According to (5.53), the policy $\mu_{\bar{k}+1}^{i\star}$ of agent $i \in \mathcal{N}$ at level $\bar{k}+1$ will satisfy, for all $s \in \mathcal{S}$,

$$
\begin{aligned}
\mu_{\bar{k}+1}^{i\star} &\in \arg\max_{\mu^i \in \mathcal{M}^i} \mathbb{E}\left[J^i_{\mu^i,\ \mu_\kappa^{-i\star}}(s) \mid \kappa \sim \mathcal{P}_{\bar{k}+1}\right] \\
&= \arg\max_{\mu^i \in \mathcal{M}^i} \sum_{\kappa=0}^{\bar{k}-1} P_{\bar{k}+1}(\kappa) J^i_{\mu^i,\ \mu_\kappa^{-i\star}}(s) \\
&\qquad + P_{\bar{k}+1}(\bar{k}) J^i_{\mu^i,\ \mu_{\bar{k}}^{-i\star}}(s) \qquad\qquad (5.55) \\
&= \arg\max_{\mu^i \in \mathcal{M}^i} \alpha \sum_{\kappa=0}^{\bar{k}-1} P_{\bar{k}}(\kappa) J^i_{\mu^i,\ \mu_\kappa^{-i\star}}(s) \\
&\qquad + P_{\bar{k}+1}(\bar{k}) J^i_{\mu^i,\ \mu_{\bar{k}}^{-i\star}}(s),
\end{aligned}
$$

where $\alpha = \sum_{i=0}^{\bar{k}-1} g(i)/\sum_{j=0}^{\bar{k}} g(j)$. Hence (5.55) yields

$$\begin{aligned}\mu_{\bar{k}+1}^{i\star} \in \arg\max_{\mu^i \in \mathcal{M}^i} \alpha \mathbb{E}\left[J^i_{\mu^i,\ \mu_\kappa^{-i\star}}(s) \mid \kappa \sim \mathcal{P}_{\bar{k}}\right] \\ + P_{\bar{k}+1}(\bar{k}) J^i_{\mu^i,\ \mu_{\bar{k}}^{-i\star}}(s),\ s \in \mathcal{S}.\end{aligned} \tag{5.56}$$

However, by the definition of the cognitive hierarchy model, $\mu_{\bar{k}}^{i\star}$ maximizes $\mathbb{E}[J^i_{\mu^i,\ \mu_\kappa^{-i\star}}(s) \mid \kappa \sim \mathcal{P}_{\bar{k}}]$ with respect to $\mu^i \in \mathcal{M}^i$ for all $s \in \mathcal{S}$. In addition, since the policy tuple $\{\mu_{\bar{k}}^{i\star},\ \mu_{\bar{k}}^{-i\star}\}$ forms a Nash equilibrium, $\mu_{\bar{k}}^{i\star}$ also maximizes $J^i_{\mu^i,\ \mu_{\bar{k}}^{-i\star}}(s)$ with respect to $\mu^i \in \mathcal{M}^i$ for all $s \in \mathcal{S}$. Since the right-hand side of (5.56) is a convex combination of $\mathbb{E}[J^i_{\mu^i,\ \mu_\kappa^{-i\star}}(s) \mid \kappa \sim \mathcal{P}_{\bar{k}}]$ and $J^i_{\mu^i,\ \mu_{\bar{k}}^{-i\star}}(s)$, we conclude that $\mu_{\bar{k}+1}^{i\star} \equiv \mu_{\bar{k}}^{i\star}\ \forall i \in \mathcal{N}$. Hence, since the policy tuples $\{\mu_{\bar{k}}^{i\star},\ \mu_{\bar{k}}^{-i\star}\}$ form a Nash equilibrium, then the policy tuples $\{\mu_{\bar{k}+1}^{i\star},\ \mu_{\bar{k}+1}^{-i\star}\}$ will also form a Nash equilibrium. The result can be generalized by using induction, so that $\{\mu_k^{i\star},\ \mu_k^{-i\star}\}$ are Nash equilibria for all $k \in \mathbb{N}_{k \geq \bar{k}}$. □

Remark 5.11. As $\lambda \to \infty$ in the Poisson distribution (5.52), we have $P_k(k-1) \to 1$ and $P_k(m) \to 0$ for all $m \in \{0, 1, \ldots, k-2\}$. Hence, as $\lambda \to \infty$, the cognitive hierarchy model reduces to the level-k thinking model, because a level-k agent will believe that the rest of the agents have a cognitive level of $k-1$ with probability 1. As such, the results of Theorem 5.6 can be extended to the level-k thinking case. Notably, the same reasoning does not also apply to Theorem 5.5; if $\lambda \to \infty$, then the limit $\lim_{\kappa \to \infty} \lambda/(\kappa + 1)$ in the proof of the theorem will be indefinite. ■

5.2.6 Boundedly rational policies with limited communication

Using the foregoing analysis, only time-triggered policies can be derived; that is, every player updates their action in each time step of the stochastic game, and an infinite amount of communication resources is assumed to be available. However, communication bandwidth is often limited; this is especially the case in cyber-physical systems, where multiple physical and virtual components compete with one another to gain access to the communication channels of the system. As such, in this section, we propose an intermittent version of the boundedly rational policies presented previously, so that communication constraints are taken into account. We will consider two different approaches.

Concurrent estimation of boundedly rational policies with intermittency

In the first approach, we can design an intermittent rule that will be incorporated within the models of level-k thinking and cognitive hierarchy. As such, at each level k of level-k thinking, an intermittent policy that is the best response to an intermittent level $k-1$ policy is designed. Similarly for cognitive hierarchy, at each level k, an intermittent policy that is the best response to a distribution of lower-level intermittent policies is constructed. In short, when computing their intermittent policies, boundedly rational agents of any level take into account the fact that the other agents *also* use intermittency. In what follows, we focus on derivations for the level-k thinking model, as the results are similar for the cognitive hierarchy case.

Toward designing the aforementioned policies, we define an augmented state space of the form $\tilde{\mathcal{S}} = \mathcal{S} \times \mathcal{A}$ (Baumann et al., 2018). This set includes the nominal states $s_t \in \mathcal{S}$, along with the actions $a_{t-1} \in \mathcal{A}$ played by each player in the previous time step $t-1 \in \mathbb{N}$, creating a pair $x_t = (s_t, a_{t-1}) \in \tilde{\mathcal{S}}$. The tuple of the stochastic game is also redefined into $(\tilde{\mathcal{S}}, \mathcal{N}, \mathcal{A}, \tilde{r}, \tilde{p}, \gamma)$, so that the reward $\tilde{r}$ and the transition probabilities are defined over $\tilde{\mathcal{S}}$. We additionally denote by $\tilde{\mathcal{M}}^i$ the set of policies of agent $i \in \mathcal{N}$ over the space $\tilde{\mathcal{S}}$.

To optimize communication resources concurrently with the rewards r^i, we construct the augmented reward function $\tilde{r}^i$ for agent $i \in \mathcal{N}$ by adjoining a communication reward to r^i, so that

$$\tilde{r}^i(x_{t+1}, a_t^i, \tilde{\pi}^{-i}(x_t), x_t) = \rho^i r^i(s_{t+1}, a_t^i, \tilde{\pi}^{-i}(x_t), s_t) + (1 - \rho^i) I_{a_t^i = a_{t-1}^i}, \quad (5.57)$$

where $x_{t+1}, x_t \in \tilde{\mathcal{S}}$ are the augmented states at times $t+1$ and t, $a_t^i \in \mathcal{A}^i$ is the action of player i at time t, $\tilde{\pi}^{-i} \in \tilde{\mathcal{M}}^{-i}$ are joint policies of players in $\mathcal{N} \backslash \{i\}$, and $\rho^i \in [0, 1]$. Evidently, the reward function (5.57) is a convex combination of the original reward r^i and the indicator function $I_{a_t^i = a_{t-1}^i}$. The latter term forces constant policies to be more favorable, which subsequently reduces the communication burden of player i. The constant ρ^i is a measure of the communication capabilities of player i; if $\rho^i = 1$, then player i has infinite communication resources and is not penalized for updating their action at each time step; whereas if $\rho^i = 0$, then player i has zero bandwidth and will be rewarded only if they do not update their action. In a realistic scenario, ρ^i will take a value between zero and unity.

Over the new state space sequence $\{x_t\}_{t \in \mathbb{N}}$, considering the reward function (5.57) augmented with a communication penalty, the cumulative reward is

$$\tilde{J}^i_{\tilde{\pi}^i,\tilde{\pi}^{-i}}(x) = \mathbb{E}_{\tilde{p}}\left[\sum_{t=0}^{\infty}\gamma^t \tilde{r}^i(x_{t+1},\tilde{\pi}^i(x_t),\tilde{\pi}^{-i}(x_t),x_t)\,\middle|\, x_0 = x\right], \tag{5.58}$$

where $x_t \in \tilde{\mathcal{S}}$. Hence, following the same reasoning as in the time-triggered case, for all $k \in \mathbb{N}_+$, the level-k policy of agent $i \in \mathcal{N}$ with incorporated intermittency can be obtained as

$$\tilde{\pi}^{i\star}_k \in \arg\max_{\tilde{\pi}^i \in \tilde{\mathcal{M}}^i} \tilde{J}^i_{\tilde{\pi}^i,\ \tilde{\pi}^{-i\star}_{k-1}}(x), \quad x \in \tilde{\mathcal{S}}. \tag{5.59}$$

Notice that, unlike (5.39), in (5.59) the value function is defined over the sum of the rewards (5.57), which are augmented with a communication reward. Accordingly, the base level-0 policy of the time-triggered case (5.38) can be generalized to the intermittent case, so that for every agent $i \in \mathcal{N}$, it will be defined as

$$\tilde{\pi}^{i\star}_0(x_t) = \tilde{\pi}^{i\star}_0(s_t,\ a_{t-1}) = a^i_{t-1},\ a^i_{t-1} \in \mathcal{A}^i,\ s_t \in \mathcal{S}. \tag{5.60}$$

Essentially, a level-0 agent's policy is constant, as they will always play the action used in a previous time step. Notice that as (5.59)–(5.60) are of the same form as (5.38)–(5.39), Algorithms 5.2 and 5.3 can be effectively utilized to find the analytic form of the event triggered level-k policies $\tilde{\pi}^{i\star}_k$, $i \in \mathcal{N}$, $k \in \mathbb{N}$.

Although Eqs. (5.59)–(5.60) describe an effective way of optimally incorporating communication constraints within the level-k thinking model, they suffer from a drawback; information regarding the actual best-response policies is diluted by the intermittency as the levels of intelligence increase and depends on the values of ρ^i, $i \in \mathcal{N}$. This is because intermittency "quantizes" a player's policy, making it piecewise constant to save communication resources. In fact, for certain values of ρ^i, a level-k agent becomes indistinguishable from a naive level-0 agent. This is illustrated in the following theorem.

Theorem 5.7. *Consider an agent $i \in \mathcal{N}$ with a level of intelligence $\tilde{k} \in \mathbb{N}_+$, who chooses their intermittent boundedly rational policy $\tilde{\pi}^{i\star}_{\tilde{k}}$ according to recursion (5.59)–(5.60). Then there exists $\rho^\star \in (0,\ 1)$ such that for all $\rho^i < \rho^\star$, we have that $\tilde{\pi}^{i\star}_{\tilde{k}} = \tilde{\pi}^{i\star}_0$.*

Proof. Let $\rho^i = 0$. Then the value function (5.58) takes the form

$$\tilde{J}^i_{\tilde{\pi}^i,\tilde{\pi}^{-i}}(x) = \mathbb{E}_{\tilde{p}}\left[\sum_{t=0}^{\infty} \gamma^t I_{\tilde{\pi}^i(x_t)=a^i_{t-1}} \,\middle|\, x_0 = x\right].$$

Hence by (5.59) we have

$$\tilde{\pi}^{i\star}_{\tilde{k}} \in \arg\max_{\tilde{\pi}^i \in \tilde{\mathcal{M}}^i} \mathbb{E}_{\tilde{p}}\left[\sum_{t=0}^{\infty} \gamma^t I_{\tilde{\pi}^i(x_t)=a^i_{t-1}} \,\middle|\, x_0 = x\right], \quad x \in \tilde{\mathcal{S}},$$

which, since $x_t = (s_t, \ a_{t-1})$, yields

$$\tilde{\pi}^{i\star}_{\tilde{k}}(x_t) = \tilde{\pi}^{i\star}_{\tilde{k}}(s_t, a_{t-1}) = a^i_{t-1}, \ s_t \in \mathcal{S}. \tag{5.61}$$

Therefore by (5.60) and (5.61) we derive $\tilde{\pi}^{i\star}_{\tilde{k}} = \tilde{\pi}^{i\star}_0$ if $\rho^i = 0$. Finally, by the continuity of $\tilde{r}^i$ with respect to ρ^i we conclude that there exists $\rho^\star \in (0, 1)$ such that $\tilde{\pi}^{i\star}_{\tilde{k}} = \tilde{\pi}^{i\star}_0$ for all $\rho^i < \rho^\star$. □

Sequential estimation of boundedly rational policies with intermittency

In the second approach, we can design the intermittency rule a posteriori, so that it is distinct from the models of bounded rationality. In particular, the level-k policies are initially derived as in the time-triggered case, and they are subsequently "quantized" to obtain their intermittent version. As a result, although each level-k agent can follow an intermittency scheme a posteriori, they do not assume that the lower-level agents do so as well. This approach can solve the problem of information dilution described previously, but it has a different drawback: it leads to overall suboptimality, as the decision of whether to communicate or not is taken *after* the level-k policies have been designed.

Specifically, given the nominal level-k policies (5.38)–(5.39) for player $i \in \mathcal{N}$, a communication aware policy $\hat{\pi}^i_k$ at the level $k \in \mathbb{N}_+$ can be designed in an a posteriori sense, so that

$$\hat{\pi}^i_k(x_k) = \begin{cases} a^i_{t-1}, & \gamma^i_k(x_k) = 0, \\ \pi^{i\star}_k(s_k), & \gamma^i_k(x_k) = 1. \end{cases} \tag{5.62}$$

In (5.62), $\gamma^i_k \in \Gamma := \tilde{\mathcal{S}} \to \{0, 1\}$ is an event that indicates whether communication will take place or not, and which should be optimized. In that sense, its optimal value $\gamma_k^{i\star}$ should be such that

$$\gamma_k^{i\star} \in \arg\max_{\gamma_k^i \in \Gamma} \hat{J}^i_{\hat{\pi}_k^i,\ \pi_{k-1}^{-i\star}}(x), \quad x \in \tilde{S}, \tag{5.63}$$

where

$$\hat{J}^i_{\hat{\pi}_k^i, \pi_{k-1}^{-i\star}}(x) = \mathbb{E}_{\tilde{p}}\left[\sum_{t=0}^{\infty} \gamma^t \tilde{r}^i(x_{t+1}, \hat{\pi}_k^i(x_t), \pi_{k-1}^{-i\star}(s_t), x_t) \,\middle|\, x_0 = x \right].$$

Hence, given (5.62) and (5.63), the optimal intermittent policy, designed in a sequential, a posteriori manner, is given by

$$\hat{\pi}_k^{i\star}(x_k) = \begin{cases} a_{t-1}^i, & \gamma_k^{i\star}(x_k) = 0, \\ \pi_k^{i\star}(s_k), & \gamma_k^{i\star}(x_k) = 1. \end{cases}$$

Remark 5.12. Optimization (5.63) can be performed using standard policy or value iteration techniques. ■

5.2.7 Simulation results

We consider a pursuit-evasion game, taking place in the grid world depicted in Fig. 5.7. There are two players participating in the game, the pursuer (player P) and the evader (player E). The goal of the pursuer is to eventually be positioned in the same tile as the evader, whereas the evader wants to avoid such a situation. To impose the aforementioned behavior, the pursuer is given a reward of 10 when they are positioned in the same tile as the evader, whereas the evader takes a reward of −10. In every other case, both the pursuer and the evader are given a reward equal to 0.

The action set of both the pursuer and the evader consists of the actions {Right, Up, Left, Down, Stay}, which indicate the wish of the corresponding agent to move one tile right, up, left, down, or not move at all, respectively. Such a wish will be fulfilled with probability equal to 0.92, whereas the agent will be moved to an arbitrary other neighboring tile with probability 0.08 (0.02 for each other tile). In case that a particular transition is not possible due to being positioned at the border of the grid or due to an obstacle, the transition probabilities are normalized accordingly. Finally, if the pursuer is positioned in the same tile as the evader, then the probability of both agents remaining in the same tile is set to 0.8 no matter which actions are chosen; as such, the evader gets only a slim chance to escape once caught.

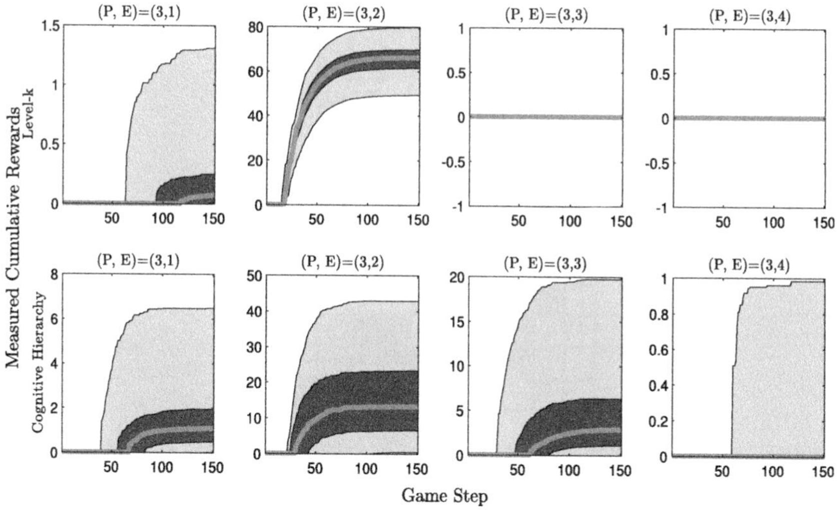

Figure 5.6 Evolution of the cumulative reward of a level-3 pursuer over 1000 sampled games, where the level of the evader varies from 1 to 4. The median is depicted in red (mid gray in print version), the $\pm 20\%$ interval in blue (dark gray in print version), and the $\pm 40\%$ interval in cyan (light gray in print version).

First, we consider the level-k thinking and the cognitive hierarchy models, with $\lambda = 2$, for the time-triggered scenario. We assume that the level of the pursuer is 3, whereas the level of the evader is 1, 2, 3, or 4. We choose the level-0 policy to be constant and always equal to "Right". The discount factor for the rewards is set equal to $\gamma = 0.95$.

Up to 1000 different evolutions of the game are sampled, given the aforementioned specifications. The initial state in these games is such that the pursuer is positioned in tile (5, 5), and the evader in tile (5, 10). The results for the first 150 steps of each game can be seen in Fig. 5.6. As expected, in level-k thinking, a level-3 pursuer responds extraordinarily well against a level-2 evader; the median number of steps needed before catching the evader is significantly small. Nevertheless, this level-k thinking policy is not robust if the level of the evader is not equal to 2. Particularly, the evader is almost never caught if its level is 3 or 4, and even a level-1 evader cannot be pursued easily. The cognitive hierarchy model, on the other hand, presents more robustness: although the average cumulative reward of the pursuer against a level-2 evader is not as high, the cumulative reward is consistently better across the rest of the evader's levels. This is expected, as

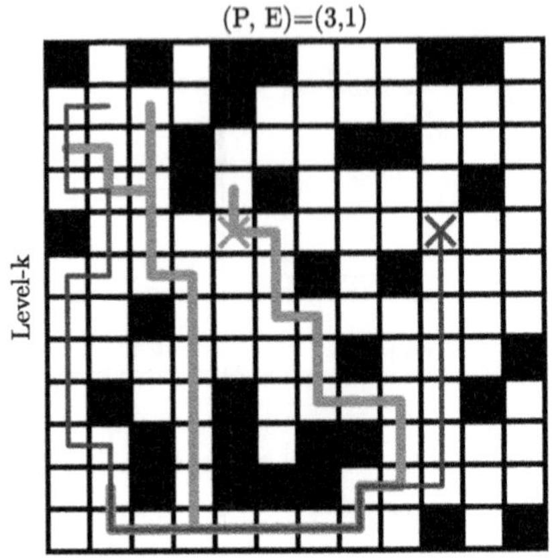

Figure 5.7 Level-3 pursuer [red (mid gray in print version)] vs. level-1 evader [blue (dark gray in print version)] in the level-*k* thinking scenario.

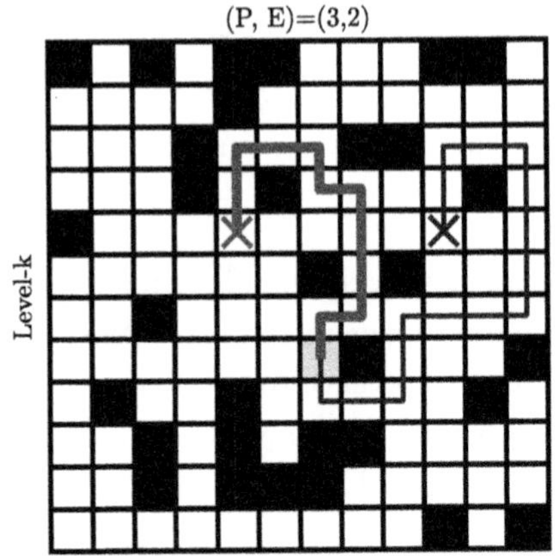

Figure 5.8 Level-3 pursuer [red (mid gray in print version)] vs. level-2 evader [blue (dark gray in print version)] in the level-*k* thinking scenario.

an agent following a cognitive hierarchy model of rationality assumes that the rest of the agents' levels follow a distribution and are not deterministic.

The level-*k* and cognitive hierarchy models are further depicted in Figs. 5.7–5.14, where the paths taken in the game closest to the median of the 1000 sampled games are illustrated (up to 70 steps of the paths, or until the evader is first caught). We can see that in both models of hierarchy, a level-2 evader is easily caught by a level-3 pursuer. However, if the evader is level 4, the task of the pursuer becomes more difficult. In fact, in the level-*k* thinking scenario, the level-4 evader easily fools the pursuer with

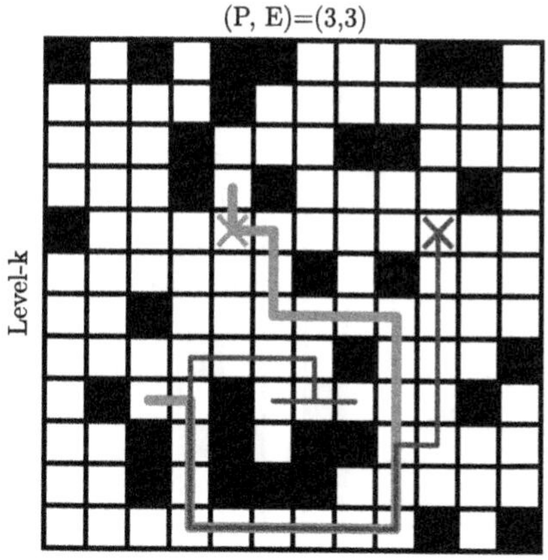

Figure 5.9 Level-3 pursuer [red (mid gray in print version)] vs. level-3 evader [blue (dark gray in print version)] in the level-*k* thinking scenario.

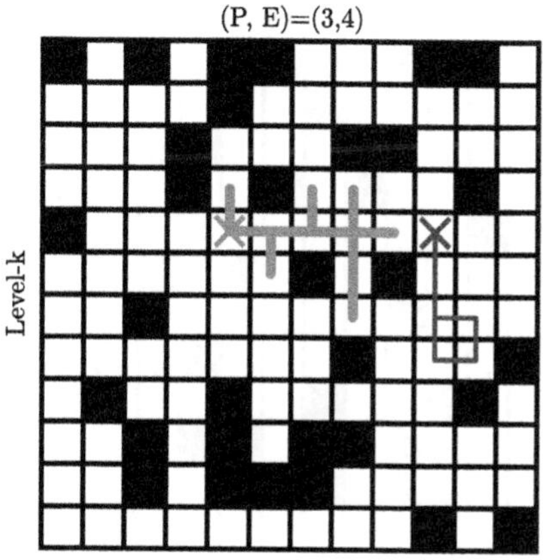

Figure 5.10 Level-3 pursuer [red (mid gray in print version)] vs. level-4 evader [blue (dark gray in print version)] in the level-*k* thinking scenario.

only little movement and without their paths ever intersecting. Additionally, we also notice that a level-3 pursuer performs well against an evader of levels 1–3 in the cognitive hierarchy scenario, which is not the case with the level-*k* thinking. Finally, it is verified that the cognitive hierarchy policies remained constant after level 4, which was expected according to Theorem 5.5.

Next, we compare the level-recursive Algorithm 5.2 with the level-paralleled Algorithm 5.3 in terms of the number of repeat-while loops executed before termination, as well as the corresponding relative compu-

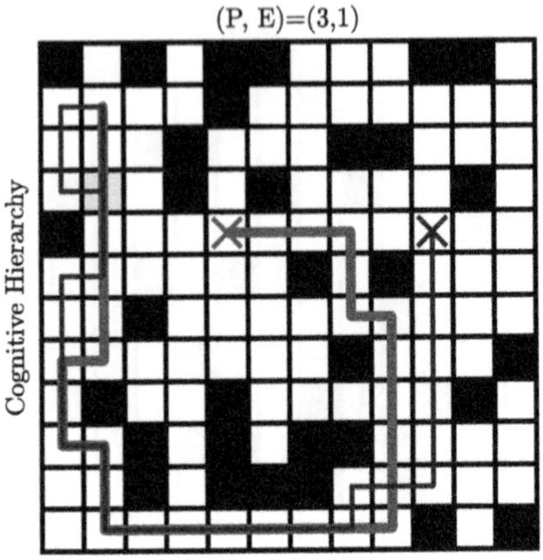

Figure 5.11 Level-3 pursuer [red (mid gray in print version)] vs. level-1 evader [blue (dark gray in print version)] in the cognitive hierarchy scenario.

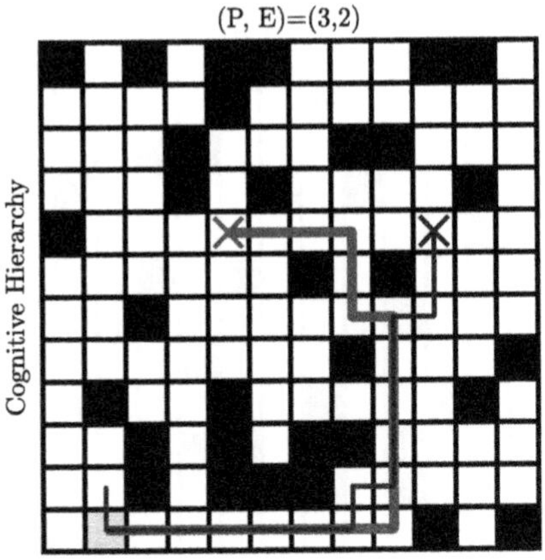

Figure 5.12 Level-3 pursuer [red (mid gray in print version)] vs. level-2 evader [blue (dark gray in print version)] in the cognitive hierarchy scenario.

tational time. Fig. 5.15 depicts the number of loop executions for both algorithms, whereas Table 5.1 shows the relative speed-up. As we can notice, the level-paralleled algorithm executes its loop significantly less before convergence. Of course, the level-paralleled repeat-while loop is more computationally intensive, relatively to the level-recursive one. However, if communication delays are low, then the asynchronous execution of the for-loops it contains (with respect to k) can offer an important speed-up, as seen in Table 5.1. Additionally, Table 5.2 shows the number of states in which the actions given by the policies derived from Algorithms 5.2 and 5.3 differ.

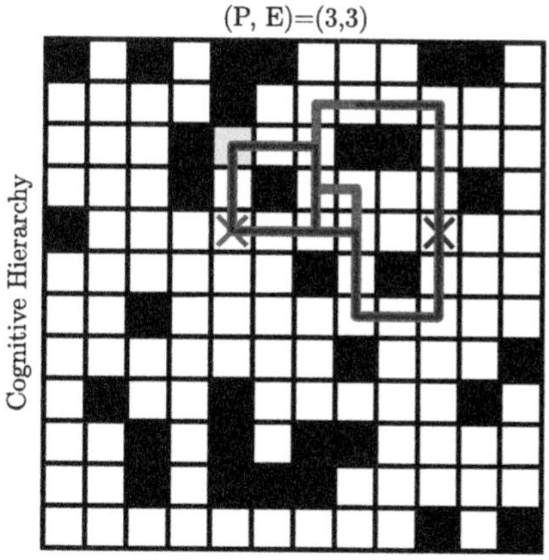

Figure 5.13 Level-3 pursuer [red (mid gray in print version)] vs. level-3 evader [blue (dark gray in print version)] in the cognitive hierarchy scenario.

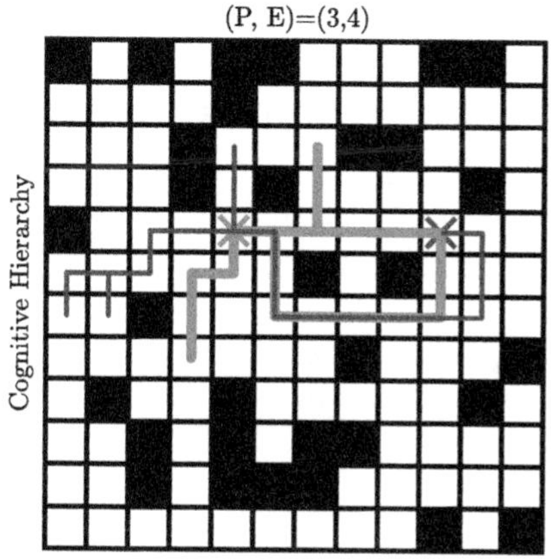

Figure 5.14 Level-3 pursuer [red (mid gray in print version)] vs. level-4 evader [blue (dark gray in print version)] in the cognitive hierarchy scenario.

As expected by Theorems 2.1 and 2.2, the discrepancies between the two policies vanish as the tolerance of the two algorithms is decreased.

Finally, we sample another 1000 games, where both the pursuer and the evader use policies that account for communication efficiency. These policies are chosen according to level-k thinking with either sequential or concurrent intermittency. We assume again that the level of the pursuer is 3, whereas the level of the evader varies from 1 to 4. In addition, we study two cases: first, when $\rho^i = 0.1$, $i = 1, 2$; and second, when $\rho^i = 1$, $i = 1, 2$. The average total number of communication instances needed up until each step

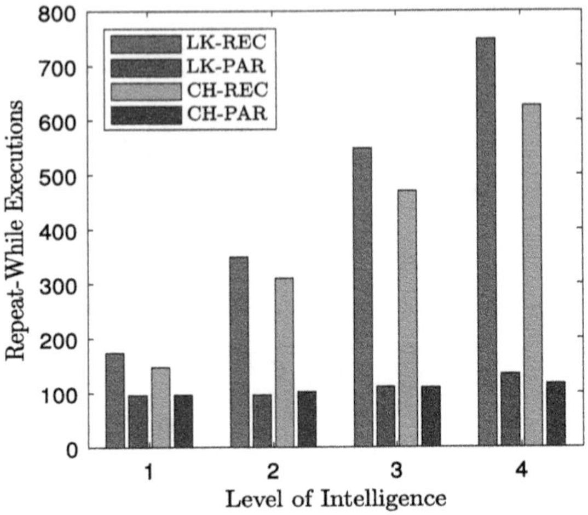

Figure 5.15 Number of repeat-while loop executions needed to find a level-*k* agent's policy with four different algorithms.

Table 5.1 Relative speedup attained with Algorithm 5.3.

Level of intelligence	**1**	**2**	**3**	**4**
Level-*k* thinking	-	1.8177	2.4685	2.7910
Cognitive hierarchy	-	1.5196	2.1364	2.6795

Table 5.2 Number of states with different actions in the policies derived by Algorithms 5.2 and 5.3 for level-*k* thinking.

Level of intelligence	**1**	**2**	**3**	**4**
$\epsilon = 10^{-2}$	15	2028	812	3413
$\epsilon = 10^{-4}$	7	30	349	31
$\epsilon = 10^{-6}$	4	4	11	8
$\epsilon = 10^{-8}$	0	0	0	0

of the game is shown in Fig. 5.16. We can notice that, in general, more communication resources are needed when concurrent event triggering is used instead of sequential triggering. This is not unexpected; in concurrent event triggering, a compromise between optimizing the communication resources and the actual objective is carried out. Hence, at instances that are crucial in optimizing the actual objective, communication is forced to take place. We also notice that while setting $\rho^i = 1$, $i = 1,\ 2$, leads to a significant increase in communication instances, communication needs not take place

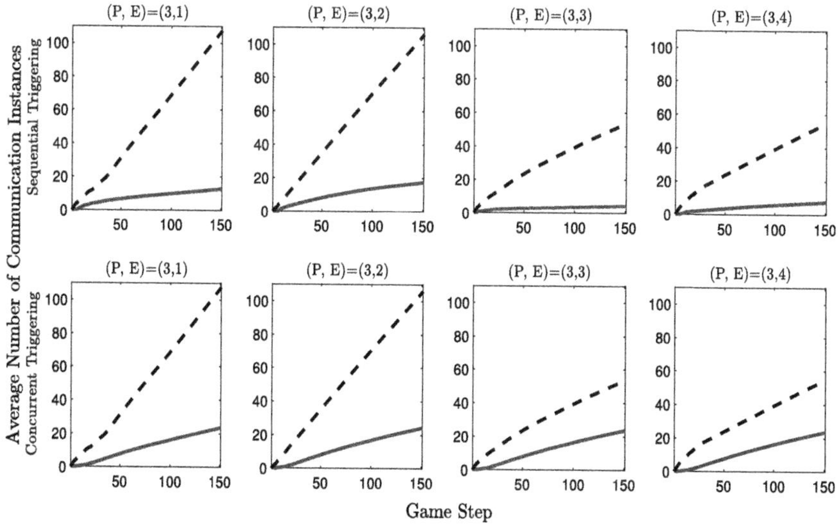

Figure 5.16 Evolution of the average total number of communication instances of the pursuer up to each step, when policies with intermittency are employed. The pursuer is a level 3 agent, while the level of the evader varies from 1 to 4. The solid lines depict the average total number of communication instances in the $\rho^1 = \rho^2 = 0.1$ case, while the dashed ones depict the $\rho^1 = \rho^2 = 1$ case.

at all times; it is only necessary when the new action at some step happens to be different from the latest action taken.

5.3. Modeling of learning attackers

In this subsection, we expand upon the previously presented ideas of bounded rationality by considering the potentially adaptive and dynamic nature of the adversaries of the CPS. We explore this issue via a framework in which an agent, whom we name "intelligent player" (IP), can learn the learning mechanism of an opponent through repeated interactions in a game scenario. In this way, we explicitly consider attackers that evolve over time, and we propose a way for the CPS operator to develop models of accurate prediction.

5.3.1 Problem formulation

Let $N \in \mathbb{N}_{\geq 2}$ be the number of players taking part in a non-cooperative game. Denote by $\mathcal{N} = \{1, \ldots, N\}$ the set of players. Each player $i \in \mathcal{N}$ has access to a finite set of actions $\mathcal{A}_i$ (the set of pure strategies) and a utility function, which she seeks to maximize. The utility function comprises a mapping from the Cartesian product of the players' action sets to the reals, i.e., $\bar{u}_i : \times_{i\in\mathcal{N}} \mathcal{A}_i \to \mathbb{R}$. Also, it is useful to define as $-i$ the set of players $j \in \mathcal{N} \setminus \{i\}$, i.e., all the players except i.

Furthermore, we allow for the inclusion of mixed strategies that assign probabilities to the actions of the player. In this case the utility function is taken to be the expected reward of the players,

$$\bar{u}_i(x) = E_{a\sim x_i}[\bar{u}_i(a)],$$

where $a \in \times_{i\in\mathcal{N}} \mathcal{A}_i$, a pure policy tuple, $x = [x_1^{\mathrm{T}}, \cdots, x_N^{\mathrm{T}}]^{\mathrm{T}} \in \times_{i\in\mathcal{N}} \Delta(\mathcal{A}_i)$, and $x_j = [x_j^1 \; x_j^2 \; \ldots \; x_j^{m_j}]$ is the point in the probability simplex embedded on $\mathcal{A}_i$ that describes the mixed strategy of player i, with $x_i^j \in [0, 1]$ the probability with which player i chooses action j. Finally, any mixed strategy tuple can be written as x or (x_i, x_{-i}), $i \in \mathcal{N}$, interchangeably.

The games considered in this work are single-shot but played repeatedly by adapting intelligent agents. Initially, we will assume that the game structure remains invariant between plays, and thus the agents learn how to play a static game in mixed strategies. We will utilize the framework introduced in Mertikopoulos and Sandholm (2018) and further investigated in Gao and Pavel (2018), where a score function that enables memory capabilities for the learning agents is employed alongside a decision rule, stemming from an optimization problem on the simplex. This framework shares similarities with Reinforcement Learning (RL) methods in MDPs and dynamical systems as well as with evolutionary game dynamics for static games of large populations.

Each player maintains a score function $z_i^j(t)$, $i \in \mathcal{N}, j \in \mathcal{A}_i$, that encodes the memory of player i about the utility gathered by playing action j. Toward this, Gao and Pavel (2018) employed an exponentially decreasing form of the gathered utility

$$z_i^j(t) = e^{-\gamma t} z_{i0}^j + \gamma \int_0^t e^{-\gamma(t-\tau)} \bar{u}_i(j, x_{-i}(\tau)) \mathrm{d}\tau, \quad t \geq 0,$$

where z_{i0}^j is the initial bias of the player toward a certain action. During the adaptation period, it will be useful to consider the vector

$$z_i(t) = [z_i^1(t) \; z_i^2(t) \; \cdots \; z_i^{\mathrm{card}(\mathcal{A}_i)}(t)]^{\mathrm{T}} \in \mathbb{R}^{\mathrm{card}(\mathcal{A}_i)}$$

as the one that is utilized in each player's learning procedure as well as the function $u_i(x(t)) : \times_{i\in\mathcal{N}} \mathcal{A}_i \to \mathbb{R}^{\mathrm{card}(\mathcal{A}_i)}$ that denotes the vector of rewards that player i would receive if they would choose different pure actions. As an example, consider a zero-sum matrix game, where $\bar{u}_i(x) = x_1^{\mathrm{T}} G x_2$ is the utility of player 1. Then $u_1(x) = G x_2$ is the vector containing the pure action rewards.

Now the evolution of the vector scoring function can be described via the dynamical system, $\forall t \geq 0$,

$$\dot{z}_i(t) = -\gamma z_i(t) + \gamma u_i(x(t)), \;\; z_i(0) = z_{i0}, \;\; t \geq 0, i \in \mathcal{N}, \tag{5.64}$$

given initial biases z_{i0}, $i \in \mathcal{N}$. The choice function used to map the score vector is derived as the solution to an optimization problem on the simplex. Along with the player's myopic objective to maximize their utility via the scores of the different actions, we can consider various cost functions that define the expected behavior of the learning agents. Gao and Pavel (2018) devise an optimization problem maximizing the expected cost $\sum_{j \in \mathcal{A}_i} x_i^j z_i^j(t)$ and minimizing the Gibbs entropy of the simplex $\sum_{j \in \mathcal{A}_i} x_i^j \log(x_i^j)$, $i \in \mathcal{N}$, thus forcing the agents to favor completely mixed strategies. A geometric interpretation of this framework is given in Mertikopoulos and Sandholm (2018). Application of this approach leads to the inclusion of a choice map such that

$$x_i(t) = \sigma_i(z_i(t)), \;\; t \geq 0, \tag{5.65}$$

where $\sigma_i : \mathbb{R}^{\mathrm{card}(\mathcal{A}_i)} \rightarrow \Delta(\mathcal{A}_i)$ maps the score vector to points in the simplex, i.e., it constitutes the solution to the optimization problem in Gao and Pavel (2018). An example of such a choice map is the softmax function given as

$$x_i^j(t) = \frac{e^{z_i^j}}{\sum_{k \in \mathcal{A}_i} e^{z_i^k}}, \;\; t \geq 0, j \in \mathcal{A}_i, \;\; i \in \mathcal{N}. \tag{5.66}$$

The presented framework constructs a continuous-time learning algorithm through which the players are able – under certain assumptions – to reach a consensus that can be shown to constitute a Nash equilibrium (Mertikopoulos et al., 2018). In this work, we bring notions of control theory to complement the increasing interest in the system-theoretic analysis of learning dynamics in games. This way, it will be possible to model and investigate complex behaviors arising in games where agents with different capabilities interact dynamically and adapt to the feedback from the game.

5.3.2 Game metalearning via cognitive heterogeneity

In this section, we introduce the algorithms which attempt to capture the heterogeneity in the cognitive abilities of competitive interacting agents. We further adopt the viewpoint of an "intelligent player" (IP) – henceforth indexed as player i – who can reason not only on the game itself, but also on the underlying learning mechanisms according to which her opponents adapt. Different from preexisting non-equilibrium game frameworks, which focus on the cognitive processes of the decision-making mechanisms, we apply bounded rationality ideas on the learning mechanisms themselves, thus claiming that a player with different cognitive abilities is not one that *plays* differently, but rather one that *learns* differently. In our case, the difference is found in the ability to perceive the learning processes of the opponents with the goal of exploiting it via various mechanisms.

Initially, we will assume that during this process, the IP is not preoccupied with exploiting the knowledge of her opponents' learning mechanism to manipulate it and maximize their own reward, but rather with inferring the ways that the players value signals from the environment and how they leverage those signals to optimize their strategies.

Toward this, the IP will need to learn the learning mechanism (5.64)–(5.65) of the rest of the players. The game is played iteratively, where the IP employs a mechanism that mimics the learning structure of the other players while it infers the way they perceive the game; their utilities $u_{-i}(x(t))$ and choice maps $\sigma_{-i}(z_{-i}(t))$, $t \geq 0$.

5.3.3 Learning the choice map

The first proposed mechanism focuses on the problem of learning the choice map $\sigma_{-i}(z_{-i}(t))$ given measurements of actions $x_{-i}(t)$ observed during play. For simplicity, we aggregate the score and decision vectors for all players $j \in \mathcal{N} \setminus \{i\}$, such that

$$
\begin{aligned}
z_{-i}(t) &\in \mathbb{R}^{\operatorname{card}\left(\times_{j\in\mathcal{N}\setminus\{i\}} \mathcal{A}_j\right)} =: \mathcal{Z}_{-i}, \\
x_{-i}(t) &= \sigma_{-i}(z_{-i}(t)) \in \bigtimes_{j\in\mathcal{N}\setminus\{i\}} \Delta(\mathcal{A}_j) =: \mathcal{D}_{-i}, \quad \forall t \geq 0.
\end{aligned}
$$

We introduce the following assumptions.

Assumption 5.5. The choice map $\sigma_{-i}(\zeta)$ can be expanded via a nonlinear function basis $\phi(\zeta) \in \mathbb{R}^k$ inside a compact set. Specifically, let $\Omega_\phi \subset \mathcal{Z}_{-i}$ be compact. Then there is $W \in \mathbb{R}^{k\times \operatorname{card}(\mathcal{D}_{-i})}$ such that $\sigma_{-i}(\zeta) = W^{\mathrm{T}}\phi(\zeta)$, $\zeta \in \Omega$. ■

Remark 5.13. For ease of exposition, we omit the inclusion of possible approximations errors, since the function we approximate and the basis used map to compact sets, and thus the approximation errors are globally bounded. Consequently, our analysis would still hold in the presence of those errors. ■

Assumption 5.6. The basis functions are bounded, i.e., there is $\phi_{\max} \in \mathbb{R}^+$ such that $\|\phi(\cdot)\| \leq \phi_{\max}$. ■

Remark 5.14. For simplicity, in this work, we assume that there is a function basis that allows us to express the choice maps of the players via a linear combination of those functions. However, differently from more common uses of approximation structures, the space over which our estimates map to is a probability simplex. Thus some of the usually employed basis functions may not satisfy this constraint. Although there are various ways, we can guarantee that during transience – e.g., through the use of projection operators during the tuning process – this is not always possible. However, during this initial work on boundedly rational learning, we will show that as time goes to infinity, the approximation structures will satisfy certain convergence guarantees. Furthermore, structures from static bounded rationality – such as level-k thinking or cognitive hierarchy strategies (Camerer, 2011) – can be used to construct an appropriate basis. Specifically, we may consider that the different elements of the function basis encode different thinking steps and the weights W express the probability that some player belongs to a specific level of thinking. Moreover, the boundedness of the basis is easily guaranteed by choosing appropriate functions, such as sigmoidal functions. ■

Now the IP employs the following model of adversarial behavior estimation:

$$\dot{\hat{z}}_{-i}(t) = -\gamma \hat{z}_{-i}(t) + \gamma u_{-i}(x(t)), \ \hat{z}(0) = \hat{z}_0, \ t \geq 0, \tag{5.67}$$

$$\hat{x}_{-i}(t) = \hat{W}^{\mathrm{T}} \phi(\hat{z}_{-i}(t)), \tag{5.68}$$

where $\hat{z}_{-i} \in \mathcal{Z}_{-i}$ is the estimated score function with initial biases given by $\hat{z}_{-i0}$. The estimated strategy is $\hat{x}_{-i} \in \mathcal{D}_{-i}$, and $\hat{W} \in \mathbb{R}^k$ is the estimate of the weights that describe the choice map.

Remark 5.15. When complete knowledge of the utility functions is assumed, the objective of the IP is to derive the choice map of the opponents, i.e., to tune the weight estimates $\hat{W}$ through estimates of the score function

$\hat{z}_i$ even during the presence of discrepancies that appear because of lack of knowledge of the true initial bias z_{i0}. ■

Defining the score vector estimation error as $\tilde{z}_{-i}(t) = z_{-i}(t) - \hat{z}_{-i}(t)$ yields the score error dynamics

$$\dot{\tilde{z}}_{-i}(t) = \dot{z}_{-i}(t) - \dot{\hat{z}}_{-i}(t) = -\gamma \tilde{z}_{-i}(t), \; t \geq 0, \tag{5.69}$$

where (5.64) and (5.67) have been used.

For the IP to infer the choice map of the other players, she utilizes a gradient descent-based adaptation rule. The estimation error is defined according to the observed actions during play, $e(t) = x_{-i}(t) - \hat{x}_{-i}(t)$. Considering the squared normed error $E(t) = \frac{1}{2}e^{\mathrm{T}}(t)e(t)$ as the objective function of the gradient descent rule yields

$$\dot{\hat{W}}(t) = -\eta_W \frac{\partial E}{\partial \hat{W}} = \eta_W \phi(\hat{z}_{-i})e^{\mathrm{T}}, \; t \geq 0, \tag{5.70}$$

where $\eta_W \in \mathbb{R}^+$ is the choice map learning rate, and, as in the sequel, we omit the explicit dependence of the variables on time for ease of exposition. Define the weight estimation error $\tilde{W}(t) = W(t) - \hat{W}(t)$, whose dynamics is given by

$$\dot{\tilde{W}}(t) = -\dot{\hat{W}}(t) = -\eta_W \phi(\hat{z}_{-i})e^{\mathrm{T}} = -\eta_W \phi(\hat{z}_{-i})\left(\phi^{\mathrm{T}}(z_{-i})W - \phi^{\mathrm{T}}(\hat{z}_{-i})\hat{W}\right),$$

where we have utilized the tuning dynamics given in (5.70). By adding and subtracting $\eta_W \phi(\hat{z}_{-i})\phi^{\mathrm{T}}(\hat{z}_{-i})W$ we get

$$\begin{aligned}\dot{\tilde{W}}(t) &= -\eta_W \phi(\hat{z}_{-i})\phi^{\mathrm{T}}(\hat{z}_{-i})\tilde{W} + \eta_W \phi(z_{-i})\phi^{\mathrm{T}}(\hat{z}_{-i})W \\ &\quad - \eta_W \phi(\hat{z}_{-i})\phi^{\mathrm{T}}(\hat{z}_{-i})W \\ &= -\eta_W \phi(\hat{z}_{-i})\phi^{\mathrm{T}}(\hat{z}_{-i})\tilde{W} + \eta_W\left(\phi(z_{-i})\phi^{\mathrm{T}}(\hat{z}_{-i}) - \phi(\hat{z}_{-i})\phi^{\mathrm{T}}(\hat{z}_{-i})\right)W.\end{aligned} \tag{5.71}$$

Now we rewrite the weight error dynamics (5.71) as a sum of nominal linear dynamics alongside an input term:

$$\dot{\tilde{W}} = F(\tilde{W}) + R, \; t \geq 0,$$

where $F(\tilde{W}) = -\eta_W \phi(\hat{z}_{-i})\phi^{\mathrm{T}}(\hat{z}_{-i})\tilde{W}$ is the nominal linear time-varying drift term, and $R = R(z_{-i}(t), \hat{z}_{-i}(t)) = \eta_W(\phi(z_{-i})\phi^{\mathrm{T}}(\hat{z}_{-i}) - \phi(\hat{z}_{-i})\phi^{\mathrm{T}}(\hat{z}_{-i}))W$

is the input term. We note that for the nominal dynamics,

$$\dot{\tilde{W}} = F(\tilde{W}),$$

the exponential stability is guaranteed if $\phi(\hat{z}_{-i})$ satisfies the persistence of excitation condition (Ioannou and Fidan, 2006). Consequently, we may invoke the converse Lyapunov theorem (Khalil and Grizzle, 2002) to show that there is $V_{\tilde{W}}(\tilde{W}, t)$, a time-varying scalar function of $\tilde{W}$ and $c_1, c_2, c_3, c_4 \in \mathbb{R}^+$, such that

$$\begin{aligned} c_1\|\tilde{W}\|^2 \leq V_{\tilde{W}}(\tilde{W}, t) \leq c_2\|\tilde{W}\|^2, \\ \frac{\partial V_{\tilde{W}}}{\partial t} + \frac{\partial V_{\tilde{W}}}{\partial \tilde{W}}F(\tilde{W}) \leq -c_3\|\tilde{W}\|^2, \\ \left\|\frac{\partial V_{\tilde{W}}}{\partial \tilde{W}}\right\| \leq c_4\|\tilde{W}\|. \end{aligned} \tag{5.72}$$

The following theorem states the asymptotic stability to the origin of the errors of the inference mechanism employed by the IP.

Theorem 5.8. *Consider a non-cooperative game played between N players, adapting their strategies according to the reinforcement learning mechanisms* (5.64)–(5.65). *Let the player i be of higher cognitive ability such that she is able to infer the learning mechanisms of her opponents via a network structure* (5.67)–(5.68) *and by employing the update law* (5.70). *Then the estimated score function error $\tilde{z}$ and the weight estimation error $\tilde{W}$ have an asymptotically stable equilibrium at the origin.*

Proof. Consider the positive definite function of $\tilde{z}_{-i}$ given as $V_z(\tilde{z}_{-i}) = \frac{1}{2\gamma}\tilde{z}_{-i}^{\mathrm{T}}\tilde{z}_{-i}$. Taking the time derivative of V_z yields

$$\dot{V}_z(\tilde{z}_{-i}) = \tilde{z}_{-i}^{\mathrm{T}}\dot{\tilde{z}}_{-i} = -\tilde{z}_{-i}^{\mathrm{T}}\tilde{z}_{-i} = -\|\tilde{z}_{-i}\|^2 < 0 \quad \forall \tilde{z}_{-i} \neq 0,$$

where (5.69) has been utilized. Thus $\tilde{z}_{-i}(t)$ converges asymptotically to the origin, i.e., $\lim_{t\to\infty} \hat{z}_{-i} \doteq z_{-i}$.

Now we take the time derivative of $V_{\tilde{W}}$, which yields

$$\dot{V}_{\tilde{W}}(\tilde{W}, t) = \frac{\partial V_{\tilde{W}}}{\partial t} + \frac{\partial V_{\tilde{W}}}{\partial \tilde{W}}\left(F(\tilde{W}) + R\right).$$

Bounding this expression utilizing (5.72), we get

$$\dot{V}_{\tilde{W}}(\tilde{W}, t) \leq -c_3\|\tilde{W}\|^2 + \left\|\frac{\partial V_{\tilde{W}}}{\partial \tilde{W}}\right\| \|R\| \leq -c_3\|\tilde{W}\|^2 + c_4\|\tilde{W}\|\|R\|.$$

Finally, we note that $\lim_{t\to\infty} R = 0$, which implies that the input terms to $\dot{\tilde{W}}$ vanish at infinity, and since the nominal dynamics are asymptotically stable, we have that $\lim_{t\to\infty} \|\tilde{W}\| = 0$. □

5.3.4 Learning inference with unknown utilities

In this section, we further extend our results to develop a framework that enables the IP to infer the learning mechanisms of the opponents without knowledge of their utilities. Our analysis will show that this extra capability, while able to increase the advantage that the IP has over her opponents, comes with weaker boundedness guarantees. Our proposed methodology considers multi-player zero-sum games with linear in the decisions utilities. Those games are named "polymatrix" games (Cai and Daskalakis, 2011). Given player $i \in \mathcal{N}$, the utility $\bar{u}_i(x_i, x_{-i})$ has the form

$$\bar{u}_i(x_i, x_{-i}) = \sum_{j \in \mathcal{N} \setminus \{i\}} x_i^{\mathrm{T}} A_{ij} x_j, \tag{5.73}$$

where $A_{ij} \in \mathbb{R}^{\mathrm{card}(\mathcal{A}_i) \times \mathrm{card}(\mathcal{A}_j)}$, $i, j \in \mathcal{N}$, is the matrix quantifying the interaction between two players. Subsequently, we can rewrite the vector utility function based on (5.73) as it appears in (5.64) in linear-in-the-parameters form. Specifically, we can see that for any player $j \in \mathcal{N}$, we have that $u_j(x) = \alpha_j^{\mathrm{T}} g_j(x)$, where

$$\alpha_j = \begin{bmatrix} A_{j1}^{\mathrm{T}} & A_{j2}^{\mathrm{T}} & \cdots & A_{jN}^{\mathrm{T}} \end{bmatrix}^{\mathrm{T}} \tag{5.74}$$

is the matrix containing all the parameters of the reward that player j obtains during play, and

$$g_j(x) = \begin{bmatrix} x_1^{\mathrm{T}} & \cdots & x_{j-1}^{\mathrm{T}} & x_{j+1}^{\mathrm{T}} & \cdots & x_N^{\mathrm{T}} \end{bmatrix}^{\mathrm{T}}$$

contains the mixed strategies of all the players $k \in \mathcal{N} \setminus \{j\}$.

Remark 5.16. A natural extension of polymatrix games defined by (5.73) is the framework of network games, in which an underlying graph structure constrains the interactions between specific selfish agents. The proposed framework can be easily extended to network games in a straightforward manner. Furthermore, modulo approximation errors due to the nonlinear nature of $u_j(x)$, $j \in \mathcal{N}$, our inference approach can be utilized in either zero-sum or non-zero-sum games with arbitrary utilities. ■

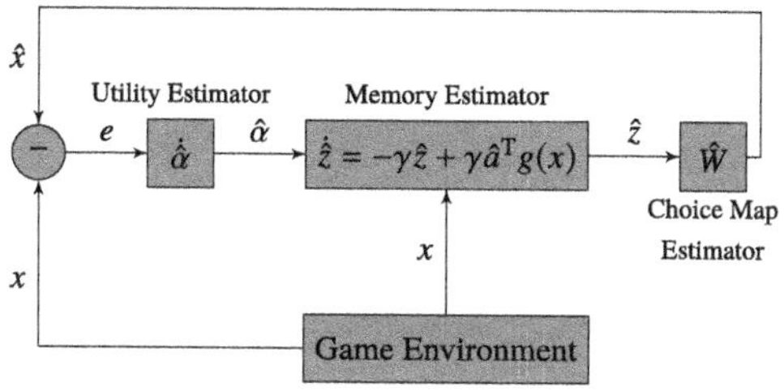

Figure 5.17 Diagram of the metalearning problem with unknown utilities.

Toward this, the IP will employ a structure that mimics (5.64) and (5.65),

$$\dot{\hat{z}}_{-i}(t) = -\gamma \hat{z}_{-i} + \gamma \hat{\alpha}_{-i}^{\mathrm{T}} g_{-i}(x),\ \hat{z}_{-i}(0) = \hat{z}_{-i0},\ t \geq 0, \tag{5.75}$$
$$\hat{x}_{-i}(t) = \hat{W}^{\mathrm{T}} \phi(\hat{z}_{-i}),$$

where $\hat{z}_{-i}$ and $\hat{x}_{-i}$ are defined as before, and $\hat{\alpha}_{-i} \in \mathbb{R}^{\mathrm{card}(Z_{-i}) \times \mathcal{D}_{-i}}$ is the estimate of the matrix defined in (5.74). We note that the difference between this problem and the one solved due to Theorem 5.8 is the existence of a utility estimation error, which leads to the score error dynamics

$$\dot{\tilde{z}}_{-i}(t) = \dot{z}_{-i}(t) - \dot{\hat{z}}_{-i}(t) = -\gamma \tilde{z}_{-i} + \gamma \tilde{\alpha}_{-i}^{\mathrm{T}} g_{-i}(x). \tag{5.76}$$

Remark 5.17. This approach requires a solution to a well-known problem in the learning literature, that of credit assignment. Specifically, the problem stems from the fact that by acquiring a single output observation – in our case $\hat{x}_{-i}$ – the intelligent agent has to infer the true values of two separate but interconnected approximation structures, the estimated utility function and the choice map. This is further exacerbated in our case due to the memory that the approximation system has via the dynamic variable $\hat{z}_{-i}$. Conceptually, as well as mathematically, our proposed framework shares similarities with the tuning process of a multi-layer recurrent neural network (see Fig. 5.17). ■

Derivation of the tuning laws that allow the IP to infer the learning mechanisms of the opponents requires leveraging the method of *backpropagation through time* (Werbos, 1990). Applying this approach to continuous-time problems makes use of *total ordered derivatives*, a mathematical tool that we present next.

"Ordered" total derivatives in continuous time: Following the work of Hanselmann et al. (2007), we employ a framework of "ordered" total derivatives in continuous time, which will allow us to derive learning rules for multi-layer recurrent approximation structures with memory. Toward this, consider a dynamical system evolving according to the differential equation $\dot{y}(t) = f(y(t); w)$, $t \geq 0$, with $y(t) \in \mathbb{R}^n$ the state variable and $w \in \mathbb{R}^l$ a parameter vector affecting the system dynamics. We seek to quantify the influence of w in the state trajectory $y(t)$. This is achieved by defining the total derivative $\mathrm{d}y/\mathrm{d}w$ and noticing that

$$\dot{y}(t) = f(y; w) \Rightarrow \frac{\mathrm{d}}{\mathrm{d}t}\frac{\mathrm{d}y}{\mathrm{d}w} = \frac{\mathrm{d}}{\mathrm{d}w}\dot{y}(y, w) = \frac{\mathrm{d}y}{\mathrm{d}w}\frac{\partial \dot{y}}{\partial y} + \frac{\partial \dot{y}}{\partial w},$$
$$\frac{\mathrm{d}}{\mathrm{d}t}\frac{\mathrm{d}y}{\mathrm{d}w} = \frac{\mathrm{d}y}{\mathrm{d}w}\frac{\partial f}{\partial y} + \frac{\partial f}{\partial w},$$

where the explicit dependence on time has been omitted for ease of exposition, and the order of differentiation can be exchanged (Hanselmann et al., 2007). Thus defining $q := (\mathrm{d}y/\mathrm{d}w)^{\mathrm{T}}$ yields, for all $t \geq 0$,

$$\dot{q}(t) = q(t)\frac{\partial f(y(t); w)}{\partial y} + \frac{\partial f(y(t); w)}{\partial w}, \quad q(0) = 0. \tag{5.77}$$

Consequently, the dynamical system (5.77) can be solved in real time to compute the total derivative.

Tuning laws without knowledge of the utility function: The estimation error measured is again given by $e(t) = x_{-i}(t) - \hat{x}_{-i}(t)$. The tuning laws are based on the normed square error of the estimation error $E(t) = \frac{1}{2}e^{\mathrm{T}}(t)e(t)$, which leads to an identical tuning law for the choice map estimator as given by (5.70) in Section 5.3.3. Similarly, we apply the tuning law for the approximation structure of the utility given by

$$\dot{\hat{\alpha}}_{-i}(t) = -\sigma\hat{\alpha}_{-i} - \eta_\alpha\frac{\partial E}{\partial \hat{\alpha}_{-i}} = -\sigma\hat{\alpha}_{-i} - \eta_\alpha e\frac{\partial e^{\mathrm{T}}}{\partial \hat{\alpha}} = -\sigma\hat{\alpha}_{-i} + \eta_\alpha e\frac{\partial \phi^{\mathrm{T}}}{\partial \hat{z}_{-i}}\hat{W}\frac{\mathrm{d}\hat{z}_{-i}^{\mathrm{T}}}{\mathrm{d}\hat{\alpha}_{-i}}. \tag{5.78}$$

Now we define the utility estimation error as $\tilde{\alpha}_{-i}(t) = \alpha_{-i}(t) - \hat{\alpha}_{-i}(t)$ with dynamics given by

$$\dot{\tilde{\alpha}}_{-i}(t) = \sigma\hat{\alpha}_{-i} - \eta_\alpha e\frac{\partial \phi^{\mathrm{T}}}{\partial \hat{z}_{-i}}\hat{W}\frac{\mathrm{d}\hat{z}_{-i}^{\mathrm{T}}}{\mathrm{d}\hat{\alpha}_{-i}}. \tag{5.79}$$

Remark 5.18. The tuning law described in (5.78) is based on the so-called σ-modification of the gradient-descent method where the damping term $\sigma\hat{\alpha}_{-i}$ is utilized to guarantee the stability of the tuning algorithm. To avoid residual estimation errors due to the damping term, there are various extensions where the term is inactive when the estimated weight is inside a predefined compact set (Ioannou and Fidan, 2006). ■

The issue that arises is due to the difficulty in computing the term $(\mathrm{d}\hat{z}_{-i}/\mathrm{d}\hat{\alpha}_{-i})^{\mathrm{T}}$, the total derivative of $\hat{z}_{-i}$ with respect to $\hat{\alpha}_{-i}$, since this dependence is derived implicitly via the dynamical equation (5.75). The approach of Hanselmann et al. (2007) circumvents this problem by leveraging "ordered" total derivatives for continuous systems.

We apply (5.77) to the dynamics of $\hat{z}_{-i}$ such that $q(t) := (\mathrm{d}\hat{z}_{-i}/\mathrm{d}\hat{\alpha}_{-i})^{\mathrm{T}} \in \mathbb{R}^{\mathrm{card}(\mathcal{Z}_{-i})\times\mathrm{card}(\mathcal{Z}_{-i})}$, which yields, for all $t \geq 0$,

$$\dot{q}(t) = q\frac{\partial \dot{\hat{z}}_{-i}}{\partial \hat{z}_{-i}} + \frac{\partial \dot{\hat{z}}_{-i}}{\partial \hat{\alpha}_{-i}} = -\gamma q + \gamma g_{-i}(x), \ \ q(0) = 0.$$

This allows us to rewrite the tuning law (5.78) and the utility error dynamics (5.79) as

$$\dot{\hat{\alpha}}_{-i}(t) = -\sigma\hat{\alpha}_{-i} + \eta_\alpha e\frac{\partial \phi^{\mathrm{T}}}{\partial \hat{z}_{-i}}\hat{W}q, \tag{5.80}$$

$$\dot{\tilde{\alpha}}_{-i}(t) = \sigma\hat{\alpha}_{-i} - \eta_\alpha e\frac{\partial \phi^{\mathrm{T}}}{\partial \hat{z}_{-i}}\hat{W}q. \tag{5.81}$$

Before stating the boundedness results of our proposed approach, the following two propositions due to Sontag (2013) are needed.

Proposition 5.1. *Consider a dynamical system evolving in* $\mathcal{M} \subseteq \mathbb{R}^l$ *according to* $\dot{y}(t) = f(y(t), t)$, $y(0) = y_0$, $t \geq 0$. *Let* $f(y, t)$ *be locally Lipschitz in* y, *continuous on* t *for all* $y(t) \in \mathcal{M}$ *and locally integrable in* t *for all* $y(t) \in \mathcal{M}$. *Then there is* $t_f \in \mathbb{R}_{>0} \cup \{\infty\}$ *such that there exists a maximal solution contained in* $\mathcal{M}$, *i.e.,* $y_m : [0, t_f) \to \mathcal{M}$. ■

Proposition 5.2. *Let Proposition* 5.1 *hold. If* $t_f < \infty$, *then for all compact* $\widetilde{\mathcal{M}} \subset \mathcal{M}$, *there is* $\tilde{t} \in [0, t_f)$ *such that* $y(\tilde{t}) \notin \widetilde{\mathcal{M}}$. ■

We are now able to state a result proving the boundedness of the error signals.

Theorem 5.9. *Consider a non-cooperative game played between* N *players, adapting their strategies according to the reinforcement learning mechanisms*

(5.64)–(5.65). *Let the player i be of higher cognitive ability such that they are able to infer the learning mechanisms of their opponents via a network structure* (5.67)–(5.68) *and by employing the update laws* (5.70) *for the choice map and* (5.80) *for the utility estimates. Finally, the dynamical system* (5.77) *is utilized to compute the effect of the memory between the two unknown structures via the total derivative* $(d\hat{z}_{-i}/d\hat{\alpha}_{-i})^T$. *Then the estimated score function error* $\tilde{z}_{-i}$, *the weight estimation error* $\tilde{W}$, *the total derivative* q, *and the utility estimation error* $\tilde{\alpha}_{-i}$ *are uniformly ultimately bounded.*

Proof. We will follow the structure of Bechlioulis and Rovithakis (2014). Our analysis involves the dynamics of the variables $\{\tilde{W},\ q,\ \tilde{\alpha}_{-i},\ \tilde{z}_{-i}\}$. We can consider the set $\mathcal{M}$ to be the Cartesian product of the spaces where the variables in the tuple lie. Thus it is both open and nonempty. Due to the structure of the overall dynamics – as given by (5.71), (5.77), (5.81), and (5.76) – and to the fact that ϕ is bounded for all arguments, since the admissible strategies evolve in a Cartesian product of compact probability simplices, the vector fields describing the evolution of the systems are locally Lipschitz continuous in the states and continuous and locally integrable in t. Then, according to Proposition 5.1, there exists a maximal solution such that the states remain inside $\mathcal{M}$ for all $t \in [0, t_m)$ for some $t_m \in \mathbb{R}_{>0} \cup \{\infty\}$.

We will now show the boundedness of the state variables over $[0, t_m)$. Thus we consider the positive definite function

$$\mathcal{V}_1(q, \tilde{W}, t) = \mathrm{tr}\left(\frac{1}{2\gamma} q^{\mathrm{T}} q\right) + V_{\tilde{W}},$$

where $V_{\tilde{W}}$ is defined in accordance to the converse Lyapunov theorem, i.e., satisfying (5.72). Taking the time derivative of $\mathcal{V}_1$ along the trajectories of q and $\tilde{W}$ yields

$$\begin{aligned}
\dot{\mathcal{V}}_1(q, \tilde{W}, t) &= \frac{1}{\gamma}\mathrm{tr}(q^{\mathrm{T}}\dot{q}) + \frac{\partial V_{\tilde{W}}}{\partial t} + \frac{\partial V_{\tilde{W}}}{\partial \tilde{W}}(F(\tilde{W})) \\
&\quad + \frac{\partial V_{\tilde{W}}}{\partial \tilde{W}}\big(\eta_W(\phi(z_{-i})\phi^{\mathrm{T}}(\hat{z}_{-i}) - \phi(\hat{z}_{-i})\phi^{\mathrm{T}}(\hat{z}_{-i}))W\big) \\
&= -\|q\|^2 + \mathrm{tr}(q^{\mathrm{T}} g_{-i}(x)) + \frac{\partial V_{\tilde{W}}}{\partial t} + \frac{\partial V_{\tilde{W}}}{\partial \tilde{W}}(F(\tilde{W})) \\
&\quad + \frac{\partial V_{\tilde{W}}}{\partial \tilde{W}}\big(\eta_W(\phi(z_{-i})\phi^{\mathrm{T}}(\hat{z}_{-i}) - \phi(\hat{z}_{-i})\phi^{\mathrm{T}}(\hat{z}_{-i}))W\big),
\end{aligned}$$

implying that $\dot{\mathcal{V}}_1$ is bounded according to

$$\dot{\mathcal{V}}_1(q, \tilde{W}, t) \leq -\|q\|^2 + \|g_{-i}(x)\|\|q\| - c_3\|\tilde{W}\|^2 + c_4\kappa\|W\|\|\tilde{W}\|,$$

where

$$\kappa := \max_{z_{-i}, \hat{z}_{-i}} \|\phi(z_{-i})\phi^{\mathrm{T}}(\hat{z}_{-i}) - \phi(\hat{z}_{-i})\phi^{\mathrm{T}}(\hat{z}_{-i})\| = 2\phi_{\max}^2.$$

The Young's inequality yields

$$\dot{\mathcal{V}}_1(q, \tilde{W}, t) \leq -\frac{1}{2}\|q\|^2 - \frac{c_3}{2}\|\tilde{W}\|^2 + \frac{1}{2}\|g_{-i}(x)\|^2 + \frac{c_4^2\kappa^2}{2c_3}\|W\|^2.$$

Thus since there is a constant $g_{-i,\max}$ such that $\|g_{-i}(x)\| \leq g_{-i,\max}$, the variables $q(t)$ and $\tilde{W}(t)$ are constrained for all $t \in [0, t_m)$ in the compact set

$$\begin{aligned}\Omega_1 = \{(\tilde{W}, q) \mid \{\mathcal{V}_1 \leq \mathcal{V}_1(0)\} \cup \{\frac{1}{2}\min(1, c_3)(\|q\|^2 + \|\tilde{W}\|^2 \\ \leq \frac{1}{2}g_{-i,\max}^2 + \frac{c_4^2\kappa^2}{2c_3}\|W\|^2\}\}.\end{aligned}$$

Subsequently, consider the positive definite function

$$\mathcal{V}_2(\tilde{\alpha}_{-i}) = \mathrm{tr}\left(\frac{1}{2\eta_\alpha}\tilde{\alpha}_{-i}^{\mathrm{T}}\tilde{\alpha}_{-i}\right)$$

and differentiate it with respect to time to get, for all $t \in [0, t_m)$,

$$\begin{aligned}\dot{\mathcal{V}}_2(\tilde{\alpha}_{-i}) &= \mathrm{tr}(\tilde{\alpha}_{-i}^{\mathrm{T}}\dot{\tilde{\alpha}}_{-i}) \\ &= \mathrm{tr}\left(\tilde{a}_{-i}^{\mathrm{T}}e\frac{\partial\phi^{\mathrm{T}}}{\partial\hat{z}_{-i}}\hat{W}q - \tilde{\alpha}_{-i}^{\mathrm{T}}\tilde{\alpha}_{-i} + \tilde{\alpha}_{-i}^{\mathrm{T}}\alpha_{-i}\right).\end{aligned} \tag{5.82}$$

Bounding (5.82) yields

$$\begin{aligned}\dot{\mathcal{V}}_2(\tilde{\alpha}_{-i}) &\leq -\|\tilde{\alpha}_{-i}\|^2 + \|\tilde{\alpha}_{-i}\|\|\alpha_{-i}\| + \|\tilde{\alpha}_{-i}\|\|\mathcal{F}\| \\ &= \|\tilde{\alpha}_{-i}\|\big(\|\alpha_{-i}\| + \|\mathcal{F}\| - \|\tilde{\alpha}_{-i}\|\big),\end{aligned}$$

where $\|\mathcal{F}\| = \max_{(\hat{z}_{-i}, z_{-i})}\|e\frac{\partial\phi^{\mathrm{T}}}{\partial\hat{z}_{-i}}\hat{W}q\|$, which is bounded due to Assumption 5.6 for the basis functions ϕ and the boundedness of $\tilde{W}$ and q, shown in (5.82) for all $t \in [0, t_m)$. This, in turn, implies that the variable $\tilde{\alpha}_{-i}$ remains in the compact set $\Omega_2 = \{\tilde{\alpha}_{-i} \mid \{\mathcal{V}_2 \leq \mathcal{V}_2(0)\} \cup \{\|\tilde{\alpha}_{-i}\| \leq \|\alpha_{-i}\| + \|\mathcal{F}\|\}\}$ for all $t \in [0, t_m)$.

Finally, consider the positive definite function $\mathcal{V}_3(\tilde{z}_{-i}) = \frac{1}{2\gamma}\tilde{z}_{-i}^{\mathrm{T}}\tilde{z}_{-i}$, and take its derivative with respect to time to get, for all $t \in [0, t_m)$,

$$\dot{\mathcal{V}}_3(\tilde{z}_{-i}) = -\tilde{z}_{-i}^{\mathrm{T}}\tilde{z}_{-i} + \tilde{z}_{-i}^{\mathrm{T}}\tilde{\alpha}_{-i}^{\mathrm{T}}g_{-i}(x),$$

which can be upper bounded as

$$\begin{aligned}\dot{\mathcal{V}}_3(\tilde{z}_{-i}) &\leq -\|\tilde{z}_{-i}\|^2 + \|\tilde{\alpha}_{-i}^{\mathrm{T}}g_{-i}(x)\|\,\|\tilde{z}_{-i}\| \\ &= \|\tilde{z}_{-i}\|\big(\|\tilde{\alpha}_{-i}^{\mathrm{T}}g_{-i}(x)\| - \|\tilde{z}_{-i}\|\big).\end{aligned}$$

Due to the boundedness of $\tilde{\alpha}_{-i}$ and the fact that $g_{-i}(x) \in \times_{j\in\mathcal{N}\setminus\{i\}} \Delta(\mathcal{A}_j)$, the variable $\tilde{z}_{-i}$ can be shown to remain in the compact set $\Omega_3 = \{\tilde{z}_{-i} \mid \{\mathcal{V}_3 \leq \mathcal{V}_3(0)\} \cup \{\|\tilde{z}_{-i}\| \leq \max_{(\tilde{\alpha}_{-i}, g_{-i}(x))} \|\tilde{\alpha}_{-i}^{\mathrm{T}}g_{-i}(x)\|\}\}$ for all $t \in [0, t_m)$. We have thus shown that X remains in the compact set $\Omega = \Omega_1 \times \Omega_2 \times \Omega_3$ for all $t \in [0, t_m)$. Hence, if we assume that $t_m < \infty$, then due to Proposition 5.2, there is $\bar{t} < t_m$ such that the states are in $\mathcal{M} \setminus \Omega$ for some $t \in [\bar{t}, t_m)$, which is a contradiction. Thus $t_m = \infty$, and the variables remain bounded for all $t \geq 0$. □

5.3.5 Simulation results

We now provide simulation results, which verify our analysis.

Matrix game with known opponent's utility: We consider a two-player non-zero sum 2×2 bimatrix game, where the expected utilities in mixed strategies for the two players are given as $\bar{u}_1 = x_1^{\mathrm{T}}Ax_2$ and $\bar{u}_2 = x_1^{\mathrm{T}}Bx_2$, where the first, IP's utility matrix is given by

$$A = \begin{bmatrix} 4 & 1 \\ 3 & 2 \end{bmatrix},$$

whereas the second player's utility matrix is also $B = A$. The two players employ the game learning algorithm given by (5.64) with a forgetting factor $\gamma = 1$, and the decision-making mechanism (5.65) is computed as (5.66). To approximate the second player's utility map, the IP uses a Radial Basis Function (RBF) network with Gaussian activation functions and 961 neurons uniformly distributed over the set $[0, 3] \times [0, 3]$. The IP modifies her action with a sufficiently rich signal for the first 50 s to achieve sufficient exploration and has a learning rate of $\eta_W = 1$. From Fig. 5.18 we can see that the identification errors $e(t)$ and $\tilde{z}_{-i}(t)$ both converge to zero.

Matrix game with unknown opponent's utility: We also consider a two-player penny matching game, structured similarly to the non-zero-

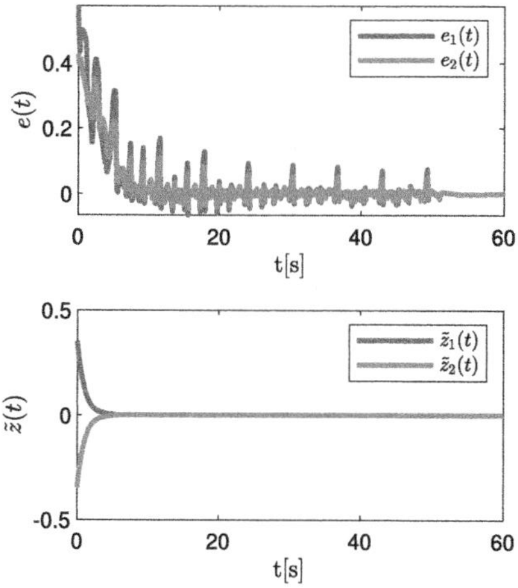

Figure 5.18 The evolution of the action identification error $e(t)$ and the gathered utility error $\tilde{z}_{-i}(t)$, where the utility matrix is known.

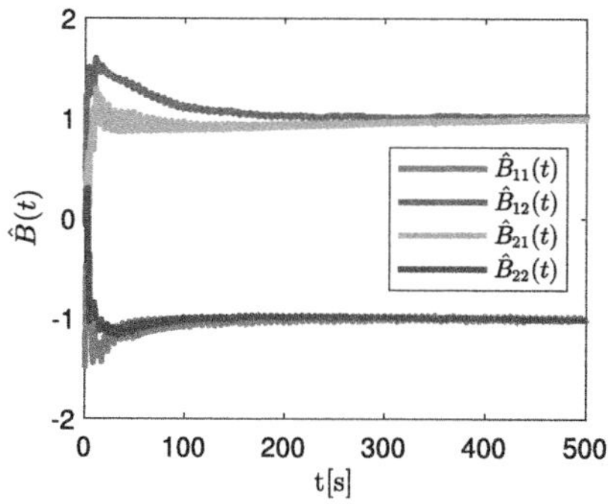

Figure 5.19 Approximation $\hat{B}$ of the utility matrix B.

sum case, where the first, IP's utility matrix is given by

$$A = \begin{bmatrix} 1 & -1 \\ -1 & 1 \end{bmatrix},$$

whereas the second player's matrix is $B = -A$, the forgetting factor in (5.64) is $\gamma = 0.65$, and the choice map is again derived as (5.66). An RBF network with Gaussian activation functions is used to approximate the second player's utility map with 121 neurons uniformly distributed over the set $[-1,\ 1] \times [-1,\ 1]$ and a tuning rate of $\eta_w = 1$. The second player's utility matrix is also approximated using (5.78) with $\eta_a = 50$. Similarly to the

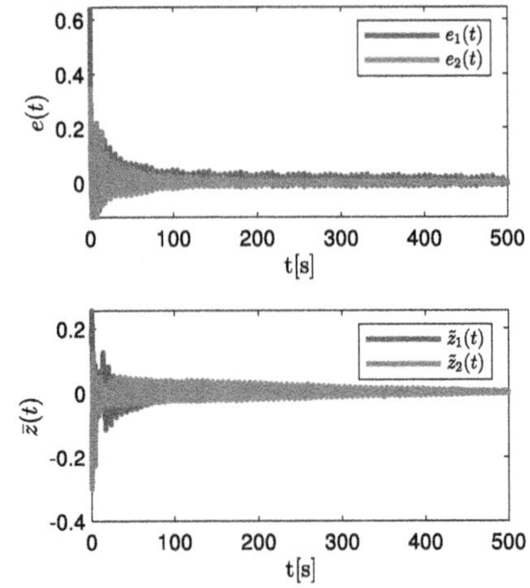

Figure 5.20 The evolution of the action identification error $e(t)$ and the gathered utility error $\tilde{z}_{-i}(t)$, where the utility matrix is unknown.

previous simulation, the IP must modify her action to achieve sufficient exploration or, equivalently, persistence of excitation.

The results are shown in Figs. 5.19 and 5.20. As we can see, the approximation $\hat{B}$ of the second player's utility matrix converges almost exactly to B. In addition, the identification errors $e(t)$ and $\tilde{z}_{-i}(t)$ converge to a very small residual set in a neighborhood of zero.

CHAPTER SIX

Future research directions

There is much work left to be done in the fields of autonomy and security for CPS. Our work has focused on bringing principles from a variety of research areas to CPS via tools from control and game theory. This approach can be expanded to further investigate notions of intelligence in adapting dynamical and learning agents interacting in highly complex and nonlinear ways.

A plethora of concepts from cyber-security can be approached from the rigorous viewpoint of control theory. Ideas of redundancy can be utilized to not only shield the systems against attackers, but also augment them with secrecy and privacy properties. Furthermore, control theorists can introduce the notions of time and dynamical behavior to security algorithms. A potential research topic stemming from our work on DoS attacks is the use of variable transmission times between the components of the CPS to increase the difficulty to the attacker. This line of research can also be explored from a decentralized control perspective, in which the plant and the controller will need to achieve synchronization in their transmission channels in a private manner.

Research on bounded rationality can lead to both the development of predictive models of behavior and to algorithms that facilitate human collaboration with CPS. Higher complexity models of bounded rationality in dynamical systems will need to consider the dynamics of learning themselves and the ways that heterogeneous learning algorithms interact. On the other hand, CPS that require humans-in-the-loop will need to be investigated in all engineering applications. These results will hopefully enable further robustification of CPS operation. As a final example, our investigation on defining computable bounded rationality strategies in stochastic games can be readily leveraged in security contexts, where appropriate underlying dynamics that describe security metrics such as those introduced in Xu (2014) are employed.

Control and Game Theoretic Methods for Cyber-Physical Security
https://doi.org/10.1016/B978-0-44-315408-9.00012-9

References

Abuzainab, N., Saad, W., Poor, H.V., 2016. Cognitive hierarchy theory for heterogeneous uplink multiple access in the internet of things. In: Information Theory (ISIT), 2016 IEEE International Symposium on. IEEE, pp. 1252–1256.

Ames, W.F., Pachpatte, B., 1997. Inequalities for Differential and Integral Equations, vol. 197. Elsevier.

An, L., Yang, G.H., 2017. Secure state estimation against sparse sensor attacks with adaptive switching mechanism. IEEE Transactions on Automatic Control.

Apostol, T.M., 1991. Calculus, vol. 1. John Wiley & Sons.

Baird, L.C., 1994. Reinforcement learning in continuous time: advantage updating. In: Neural Networks, 1994. IEEE World Congress on Computational Intelligence., 1994 IEEE International Conference on. IEEE, pp. 2448–2453.

Başar, T., Bernhard, P., 2008. H-Infinity Optimal Control and Related Minimax Design Problems: a Dynamic Game Approach. Springer Science & Business Media.

Basar, T., Olsder, G.J., 1999. Dynamic Noncooperative Game Theory, vol. 23. SIAM.

Baumann, D., Zhu, J.J., Martius, G., Trimpe, S., 2018. Deep reinforcement learning for event-triggered control. In: 2018 IEEE Conference on Decision and Control (CDC). IEEE, pp. 943–950.

Beard, R.W., McLain, T.W., 1998. Successive Galerkin approximation algorithms for nonlinear optimal and robust control. International Journal of Control 71, 717–743.

Bechlioulis, C.P., Rovithakis, G.A., 2014. A low-complexity global approximation-free control scheme with prescribed performance for unknown pure feedback systems. Automatica 50, 1217–1226.

Bertsekas, D., 2020. Multiagent value iteration algorithms in dynamic programming and reinforcement learning. Results in Control and Optimization 1, 100003.

Bertsekas, D., Tsitsiklis, J., 2015. Parallel and Distributed Computation: Numerical Methods. Athena Scientific.

Bertsekas, D.P., Tsitsiklis, J.N., 1996. Neuro-Dynamic Programming. Athena Scientific.

Bradtke, S.J., Ydstie, B.E., Barto, A.G., 1994. Adaptive linear quadratic control using policy iteration. In: Proceedings of the American Control Conference. Citeseer, p. 3475.

Brams, S.J., Kilgour, D.M., 1988. Game Theory and National Security. Blackwell.

Bronk, C., Tikk-Ringas, E., 2013. The cyber attack on Saudi Aramco. Survival 55, 81–96.

Bryson, A.E., 2018. Applied Optimal Control: Optimization, Estimation and Control. Routledge.

Cai, Y., Daskalakis, C., 2011. On minmax theorems for multiplayer games. In: Proceedings of the Twenty-Second Annual ACM-SIAM Symposium on Discrete Algorithms. SIAM, pp. 217–234.

Camerer, C.F., 2003. Behavioral game theory: plausible formal models that predict accurately. Behavioral and Brain Sciences 26, 157–158.

Camerer, C.F., 2011. Behavioral Game Theory: Experiments in Strategic Interaction. Princeton University Press.

Camerer, C.F., 2016. Behavioral game theory: psychological limits on strategic cognition. International Journal of Psychology 51, 11.

Camerer, C.F., Ho, T.H., Chong, J.K., 2004a. Behavioural game theory: thinking, learning and teaching. In: Advances in Understanding Strategic Behaviour. Springer, pp. 120–180.

Camerer, C.F., Ho, T.H., Chong, J.K., 2004b. A cognitive hierarchy model of games. The Quarterly Journal of Economics 119, 861–898.

Cao, X., Cheng, P., Chen, J., Sun, Y., 2012. An online optimization approach for control and communication codesign in networked cyber-physical systems. IEEE Transactions on Industrial Informatics 9, 439–450.

Cardenas, A.A., Amin, S., Sastry, S., 2008. Secure control: towards survivable cyber-physical systems. In: Distributed Computing Systems Workshops, 2008. ICDCS'08. 28th International Conference on. IEEE, pp. 495–500.

Casola, V., De Benedictis, A., Albanese, M., 2014. A multi-layer moving target defense approach for protecting resource-constrained distributed devices. In: Integration of Reusable Systems. Springer, pp. 299–324.

Chen, Y., Kar, S., Moura, J.M., 2016. Dynamic attack detection in cyber-physical systems with side initial state information. IEEE Transactions on Automatic Control 62, 4618–4624.

Chong, J.K., Ho, T.H., Camerer, C., 2016. A generalized cognitive hierarchy model of games. Games and Economic Behavior 99, 257–274.

Cover, T.M., Thomas, J.A., 2012. Elements of Information Theory. John Wiley & Sons.

Crawford, V.P., Iriberri, N., 2007. Level-k auctions: can a nonequilibrium model of strategic thinking explain the winner's curse and overbidding in private-value auctions? Econometrica 75, 1721–1770.

Dunlop, M., Groat, S., Urbanski, W., Marchany, R., Tront, J., 2011. MT6D: a moving target IPv6 defense. In: Military Communications Conference, 2011-Milcom 2011. IEEE, pp. 1321–1326.

Durbha, V., Balakrishnan, S., 2005. New nonlinear observer design with application to electrostatic micro-actuators. In: ASME 2005 International Mechanical Engineering Congress and Exposition. American Society of Mechanical Engineers, pp. 101–107.

Elia, N., Mitter, S.K., 2001. Stabilization of linear systems with limited information. IEEE Transactions on Automatic Control 46, 1384–1400.

Erev, I., Roth, A.E., 1998. Predicting how people play games: reinforcement learning in experimental games with unique, mixed strategy equilibria. American Economic Review, 848–881.

Farwell, J.P., Rohozinski, R., 2011. Stuxnet and the future of cyber war. Survival 53, 23–40.

Fawzi, H., Tabuada, P., Diggavi, S., 2014. Secure estimation and control for cyber-physical systems under adversarial attacks. IEEE Transactions on Automatic Control 59, 1454–1467.

Flämig, H., 2016. Autonomous vehicles and autonomous driving in freight transport. In: Autonomous Driving. Springer, pp. 365–385.

Fridman, E., Dambrine, M., 2009. Control under quantization, saturation and delay: an LMI approach. Automatica 45, 2258–2264.

Fudenberg, D., Drew, F., Levine, D.K., Levine, D.K., 1998. The Theory of Learning in Games, vol. 2. MIT Press.

Fudenberg, D., Levine, D., 1998. Learning in games. European Economic Review 42, 631–639.

Gao, B., Pavel, L., 2018. On passivity, reinforcement learning and higher-order learning in multi-agent finite games. arXiv preprint. arXiv:1808.04464.

Gao, W., Deng, C., Jiang, Y., Jiang, Z.P., 2022. Resilient reinforcement learning and robust output regulation under denial-of-service attacks. Automatica 142, 110366.

Gao, W., Jiang, Y., Jiang, Z.P., Chai, T., 2016. Output-feedback adaptive optimal control of interconnected systems based on robust adaptive dynamic programming. Automatica 72, 37–45.

Gao, W., Jiang, Z.P., 2017. Learning-based adaptive optimal tracking control of strict-feedback nonlinear systems. IEEE Transactions on Neural Networks and Learning Systems 29, 2614–2624.

Guan, Y., Ge, X., 2017. Distributed attack detection and secure estimation of networked cyber-physical systems against false data injection attacks and jamming attacks. IEEE Transactions on Signal and Information Processing over Networks 4, 48–59.

Guo, L., Liao, Q., Wei, S., Huang, Y., 2010. A kind of bicycle robot dynamic modeling and nonlinear control. In: The 2010 IEEE International Conference on Information and Automation. IEEE, pp. 1613–1617.

Hanselmann, T., Noakes, L., Zaknich, A., 2007. Continuous-time adaptive critics. IEEE Transactions on Neural Networks 18, 631–647.

He, X., Dai, H., Ning, P., 2016. Faster learning and adaptation in security games by exploiting information asymmetry. IEEE Transactions on Signal Processing 64, 3429–3443. https://doi.org/10.1109/TSP.2016.2548987.

Hespanha, J.P., 2017. Noncooperative Game Theory: An Introduction for Engineers and Computer Scientists. Princeton University Press.

Hespanha, J.P., Morse, A.S., 1999. Stability of switched systems with average dwell-time. In: Decision and Control, 1999. Proceedings of the 38th IEEE Conference on. IEEE, pp. 2655–2660.

Hewer, G., 1971. An iterative technique for the computation of the steady state gains for the discrete optimal regulator. IEEE Transactions on Automatic Control 16, 382–384.

Hoehn, A., Zhang, P., 2016. Detection of replay attacks in cyber-physical systems. In: 2016 American Control Conference (ACC). IEEE, pp. 290–295.

Hornik, K., Stinchcombe, M., White, H., 1990. Universal approximation of an unknown mapping and its derivatives using multilayer feedforward networks. Neural Networks 3, 551–560.

Huh, S., Cho, S., Kim, S., 2017. Managing IoT devices using blockchain platform. In: 2017 19th International Conference on Advanced Communication Technology (ICACT). IEEE, pp. 464–467.

Ioannou, P., Fidan, B., 2006. Adaptive Control Tutorial. SIAM.

Ioannou, P.A., Sun, J., 1996. Robust Adaptive Control, vol. 1. PTR Prentice-Hall Upper Saddle River, NJ.

Jafarian, J.H., Al-Shaer, E., Duan, Q., 2012. Openflow random host mutation: transparent moving target defense using software defined networking. In: Proceedings of the First Workshop on Hot Topics in Software Defined Networks. ACM, pp. 127–132.

Jajodia, S., Ghosh, A., Subrahmanian, V., Swarup, V., Wang, C., Wang, X., 2012. Moving Target Defense II: Application of Game Theory and Adversarial Modeling. Advances in Information Security. Springer, New York. https://books.google.com/books?id=yFzKRGJatCIC.

Jajodia, S., Ghosh, A.K., Swarup, V., Wang, C., Wang, X.S., 2011. Moving Target Defense: Creating Asymmetric Uncertainty for Cyber Threats, vol. 54. Springer Science & Business Media.

Jiang, Y., Jiang, Z.P., 2014. Robust adaptive dynamic programming and feedback stabilization of nonlinear systems. IEEE Transactions on Neural Networks and Learning Systems 25, 882–893.

Jiang, Z.P., Bian, T., Gao, W., 2020. Learning-based control: a tutorial and some recent results. Foundations and Trends® in Systems and Control 8.

Jin, X., Haddad, W.M., Jiang, Z.P., Kanellopoulos, A., Vamvoudakis, K.G., 2019. An adaptive learning and control architecture for mitigating sensor and actuator attacks in connected autonomous vehicle platoons. International Journal of Adaptive Control and Signal Processing 33, 1788–1802.

Jin, X., Haddad, W.M., Yucelen, T., 2017. An adaptive control architecture for mitigating sensor and actuator attacks in cyber-physical systems. IEEE Transactions on Automatic Control 62, 6058–6064.

Jones, A., Kong, Z., Belta, C., 2014. Anomaly detection in cyber-physical systems: a formal methods approach. In: 53rd IEEE Conference on Decision and Control. IEEE, pp. 848–853.

Khalil, H.K., Grizzle, J.W., 2002. Nonlinear Systems, vol. 3. Prentice Hall Upper Saddle River, NJ.

Kim, J., Kim, H., Lakshmanan, K., Rajkumar, R.R., 2013. Parallel scheduling for cyber-physical systems: analysis and case study on a self-driving car. In: Proceedings of the ACM/IEEE 4th International Conference on Cyber-Physical Systems. ACM, pp. 31–40.

Kiumarsi, B., Lewis, F.L., Jiang, Z.P., 2017. H∞ control of linear discrete-time systems: off-policy reinforcement learning. Automatica 78, 144–152.

Kokolakis, N.M.T., Kanellopoulos, A., Vamvoudakis, K.G., 2020. Bounded rational unmanned aerial vehicle coordination for adversarial target tracking. In: 2020 American Control Conference (ACC). IEEE, pp. 2508–2513.

Lee, E.A., 2008. Cyber physical systems: design challenges. In: 2008 11th IEEE International Symposium on Object and Component-Oriented Real-Time Distributed Computing (ISORC). IEEE, pp. 363–369.

Lee, I., Sokolsky, O., 2010. Medical cyber physical systems. In: Design Automation Conference (DAC), 2010 47th ACM/IEEE. IEEE, pp. 743–748.

Lee, R.M., Assante, M.J., Conway, T., 2014. German steel mill cyber attack. Industrial Control Systems 30, 62.

Lewis, F.L., Vrabie, D., Syrmos, V.L., 2012. Optimal Control. John Wiley & Sons.

Li, N., Oyler, D., Zhang, M., Yildiz, Y., Girard, A., Kolmanovsky, I., 2016. Hierarchical reasoning game theory based approach for evaluation and testing of autonomous vehicle control systems. In: Decision and Control (CDC), 2016 IEEE 55th Conference on. IEEE, pp. 727–733.

Liang, G., Weller, S.R., Zhao, J., Luo, F., Dong, Z.Y., 2016. The 2015 Ukraine blackout: implications for false data injection attacks. IEEE Transactions on Power Systems 32, 3317–3318.

Liberzon, D., 2012. Switching in Systems and Control. Springer Science & Business Media.

Liu, Y., Ning, P., Reiter, M.K., 2011. False data injection attacks against state estimation in electric power grids. ACM Transactions on Information and System Security 14, 13.

Lu, Z., Wang, C., Wei, M., 2016. A proactive and deceptive perspective for role detection and concealment in wireless networks. In: Cyber Deception. Springer, pp. 97–114.

Luo, B., Liu, D., Huang, T., 2014. Q-learning for optimal control of continuous-time systems. arXiv preprint. arXiv:1410.2954.

Lyshevski, S.E., 1998. Optimal control of nonlinear continuous-time systems: design of bounded controllers via generalized nonquadratic functionals. In: Proceedings of the 1998 American Control Conference. ACC (IEEE Cat. No. 98CH36207). IEEE, pp. 205–209.

Mangasarian, O.L., 1966. Sufficient conditions for the optimal control of nonlinear systems. SIAM Journal on Control 4, 139–152.

McKelvey, R.D., Palfrey, T.R., 1995. Quantal response equilibria for normal form games. Games and Economic Behavior 10, 6–38.

McLaughlin, S.E., Podkuiko, D., Delozier, A., Miadzvezhanka, S., McDaniel, P.D., 2010. Embedded firmware diversity for smart electric meters. In: HotSec.

Mertikopoulos, P., Papadimitriou, C., Piliouras, G., 2018. Cycles in adversarial regularized learning. In: Proceedings of the Twenty-Ninth Annual ACM-SIAM Symposium on Discrete Algorithms. SIAM, pp. 2703–2717.

Mertikopoulos, P., Sandholm, W.H., 2018. Riemannian game dynamics. Journal of Economic Theory 177, 315–364.

Miao, F., Zhu, Q., Pajic, M., Pappas, G.J., 2016. Coding schemes for securing cyber-physical systems against stealthy data injection attacks. IEEE Transactions on Control of Network Systems 4, 106–117.

Milošević, J., Sandberg, H., Johansson, K.H., 2018. A security index for actuators based on perfect undetectability: properties and approximation. In: 2018 56th Annual Allerton Conference on Communication, Control, and Computing (Allerton). IEEE, pp. 235–241.

Mo, Y., Kim, T.H.J., Brancik, K., Dickinson, D., Lee, H., Perrig, A., Sinopoli, B., 2012. Cyber–physical security of a smart grid infrastructure. Proceedings of the IEEE 100, 195–209.

Modares, H., Lewis, F.L., Jiang, Z.P., 2015. H_∞ tracking control of completely unknown continuous-time systems via off-policy reinforcement learning. IEEE Transactions on Neural Networks and Learning Systems 26, 2550–2562.

Molinari, B., 1976. Extended controllability and observability for linear systems. IEEE Transactions on Automatic Control 21, 136–137.

Na, J., Herrmann, G., Vamvoudakis, K.G., 2017. Adaptive optimal observer design via approximate dynamic programming. In: American Control Conference (ACC), 2017. IEEE, pp. 3288–3293.

Natarajan, B.K., 1995. Sparse approximate solutions to linear systems. SIAM Journal on Computing 24, 227–234.

Okano, K., Wakaiki, M., Yang, G., Hespanha, J.P., 2017. Stabilization of networked control systems under clock offsets and quantization. IEEE Transactions on Automatic Control 63, 1708–1723.

Okhravi, H., Hobson, T., Bigelow, D., Streilein, W., 2014. Finding focus in the blur of moving-target techniques. IEEE Security & Privacy 12, 16–26.

Pajic, M., Weimer, J., Bezzo, N., Sokolsky, O., Pappas, G.J., Lee, I., 2017. Design and implementation of attack-resilient cyberphysical systems: with a focus on attack-resilient state estimators. IEEE Control Systems 37, 66–81.

Pang, B., Bian, T., Jiang, Z.P., 2021. Robust policy iteration for continuous-time linear quadratic regulation. IEEE Transactions on Automatic Control 67, 504–511.

Pasqualetti, F., Dörfler, F., Bullo, F., 2013. Attack detection and identification in cyber-physical systems. IEEE Transactions on Automatic Control 58, 2715–2729.

Pasqualetti, F., Dorfler, F., Bullo, F., 2015. Control-theoretic methods for cyberphysical security: geometric principles for optimal cross-layer resilient control systems. IEEE Control Systems 35, 110–127.

Phillips, S., Duz, A., Pasqualetti, F., Sanfelice, R.G., 2017. Hybrid attack monitor design to detect recurrent attacks in a class of cyber-physical systems. In: 2017 IEEE 56th Annual Conference on Decision and Control (CDC). IEEE, pp. 1368–1373.

Rajkumar, R.R., Lee, I., Sha, L., Stankovic, J., 2010. Cyber-physical systems: the next computing revolution. In: Proceedings of the 47th Design Automation Conference. ACM, pp. 731–736.

Roth, A.E., Erev, I., 1995. Learning in extensive-form games: experimental data and simple dynamic models in the intermediate term. Games and Economic Behavior 8, 164–212.

Satchidanandan, B., Kumar, P.R., 2017. Dynamic watermarking: active defense of networked cyber–physical systems. Proceedings of the IEEE 105, 219–240.

Scharf, L.L., Demeure, C., 1991. Statistical Signal Processing: Detection, Estimation, and Time Series Analysis. Prentice Hall.

Shrivastava, A., Derler, P., Baboud, Y.S.L., Stanton, K., Khayatian, M., Andrade, H.A., Weiss, M., Eidson, J., Chandhoke, S., 2016. Time in cyber-physical systems. In: Proceedings of the Eleventh IEEE/ACM/IFIP International Conference on Hardware/Software Codesign and System Synthesis, pp. 1–10.

Slay, J., Miller, M., 2007. Lessons learned from the Maroochy water breach. In: International Conference on Critical Infrastructure Protection. Springer, pp. 73–82.

Song, R., Lewis, F.L., Wei, Q., Zhang, H., 2015. Off-policy actor-critic structure for optimal control of unknown systems with disturbances. IEEE Transactions on Cybernetics 46, 1041–1050.

Sontag, E.D., 2013. Mathematical Control Theory: Deterministic Finite Dimensional Systems, vol. 6. Springer Science & Business Media.

Stevens, B.L., Lewis, F.L., Johnson, E.N., 2015. Aircraft Control and Simulation: Dynamics, Controls Design, and Autonomous Systems. John Wiley & Sons.

Strzalecki, T., 2014. Depth of reasoning and higher order beliefs. Journal of Economic Behavior & Organization 108, 108–122.

Sun, Q., Zhang, K., Shi, Y., 2019. Resilient model predictive control of cyber–physical systems under dos attacks. IEEE Transactions on Industrial Informatics 16, 4920–4927.

Sutton, R.S., Barto, A.G., 1998. Reinforcement Learning: An Introduction, vol. 1. MIT Press, Cambridge.

Sutton, R.S., Barto, A.G., 2018. Reinforcement Learning: An Introduction. MIT Press.

Tambe, M., 2011. Security and Game Theory: Algorithms, Deployed Systems, Lessons Learned. Cambridge University Press.

Trentelman, H.L., Stoorvogel, A.A., Hautus, M., 2012. Control Theory for Linear Systems. Springer Science & Business Media.

Urbina, D.I., Urbina, D.I., Giraldo, J., Cardenas, A.A., Valente, J., Faisal, M., Tippenhauer, N.O., Ruths, J., Candell, R., Sandberg, H., 2016. Survey and New Directions for Physics-Based Attack Detection in Control Systems. US Department of Commerce, National Institute of Standards and Technology.

Vamvoudakis, K.G., 2017. Q-learning for continuous-time linear systems: a model-free infinite horizon optimal control approach. Systems & Control Letters 100, 14–20.

Vamvoudakis, K.G., Hespanha, J.P., Sinopoli, B., Mo, Y., 2014. Detection in adversarial environments. IEEE Transactions on Automatic Control 59, 3209–3223.

Vamvoudakis, K.G., Lewis, F.L., 2011. Multi-player non-zero-sum games: online adaptive learning solution of coupled Hamilton–Jacobi equations. Automatica 47, 1556–1569.

Vamvoudakis, K.G., Lewis, F.L., 2012. Online solution of nonlinear two-player zero-sum games using synchronous policy iteration. International Journal of Robust and Nonlinear Control 22, 1460–1483.

Vamvoudakis, K.G., Lewis, F.L., Hudas, G.R., 2012. Multi-agent differential graphical games: online adaptive learning solution for synchronization with optimality. Automatica 48, 1598–1611.

Vrabie, D., Vamvoudakis, K.G., Lewis, F.L., 2013. Optimal Adaptive Control and Differential Games by Reinforcement Learning Principles, vol. 2. IET.

Wakaiki, M., Cetinkaya, A., Ishii, H., 2019. Stabilization of networked control systems under dos attacks and output quantization. IEEE Transactions on Automatic Control 65, 3560–3575.

Wakaiki, M., Okano, K., Hespanha, J.P., 2017. Stabilization of systems with asynchronous sensors and controllers. Automatica 81, 314–321.

Weerakkody, S., Sinopoli, B., 2015. Detecting integrity attacks on control systems using a moving target approach. In: Decision and Control (CDC), 2015 IEEE 54th Annual Conference on. IEEE, pp. 5820–5826.

Werbos, P.J., 1990. Backpropagation through time: what it does and how to do it. Proceedings of the IEEE 78, 1550–1560.

Wolf, W., 2009. Cyber-physical systems. Computer 42, 88–89.

Wu, H.N., Luo, B., 2012. Neural network based online simultaneous policy update algorithm for solving the HJI equation in nonlinear H_∞ control. IEEE Transactions on Neural Networks and Learning Systems 23, 1884–1895.

Xu, S., 2014. Cybersecurity dynamics. In: Proceedings of the 2014 Symposium and Bootcamp on the Science of Security, pp. 1–2.

Yan, Y., Antsaklis, P., Gupta, V., 2017. A resilient design for cyber physical systems under attack. In: American Control Conference (ACC), 2017. IEEE, pp. 4418–4423.
Yu, X., Jiang, J., 2012. Hybrid fault-tolerant flight control system design against partial actuator failures. IEEE Transactions on Control Systems Technology 20, 871–886.
Zhai, L., Kanellopoulos, A., Fotiadis, F., Vamvoudakis, K.G., Hugues, J., 2022. Towards intelligent security for unmanned aerial vehicles: a taxonomy of attacks, faults, and detection mechanisms. In: AIAA SCITECH 2022 Forum, p. 0969.
Zhai, L., Vamvoudakis, K.G., 2021. A data-based private learning framework for enhanced security against replay attacks in cyber-physical systems. International Journal of Robust and Nonlinear Control 31, 1817–1833.
Zhai, L., Vamvoudakis, K.G., Hugues, J., 2021. Switching watermarking-based detection scheme against replay attacks, pp. 4200–4205.
Zhuang, R., DeLoach, S.A., Ou, X., 2014. Towards a theory of moving target defense. In: Proceedings of the First ACM Workshop on Moving Target Defense. ACM, pp. 31–40.

Index

Q

R

S

T

U

V

W